The Human Recip

The Human Recipe

Understanding Your Genes in Today's Society

Pascal Borry and Gert Matthijs

Leuven University Press

Original title: Iedereen geniaal. Humane genetica in woorden en cartoons

ISBN 978 94 6270 072 7
D/2016/1869/19
NUR: 870

Lay-out: Friedemann Vervoort
Cartoons: Joris Snaet

Table of Contents

Introduction

Research in the field of human genetics extends from discovery to breakthrough. It is a fascinating and stimulating discipline, but what does it mean for our society? You cannot open a newspaper or magazine without finding an article in some way related to the subject. Genetics is an important topic in everyone's lives, not just for those in the field.

This book contains a wide selection of topics, and tries to give you a clear insight into the matter, it sketches scientific, clinical and societal perspectives on the issues. A number of stories will be told: Angelina Jolie trying to raise awareness of hereditary breast cancer, the mapping of a Neanderthal's genome, President Obama's DNA for sale on eBay, a DNA test that is praised for identifying sports talent, a new method of testing that will profoundly change prenatal screening, and more.

Genetics is getting closer and closer to our own lives, not just because of these facts and stories, but also because it is inherent in each and every one of us. We all have genes and chromosomes but do not fully understand what it all means. What can the genome tell us? What is it hiding from us? What should everyone know about their DNA? Who has access to genetic data? And how will genetics influence our healthcare?

To facilitate the process of explaining human genetics and its societal and ethical questions, a skillful cartoonist has helped us out. Through edgy drawings and surprising perspectives, Joris Snaet manages to illustrate these complex themes and make them easier to understand. Using humor, human genetics becomes accessible to everyone.

The Human Recipe does not give an all-inclusive oversight of human genetics; it is a book in which heredity is linked to the possibilities and challenges of knowing our individual genome today. The book doesn't focus on describing the technical details, but engages with the increased societal challenges offered by scientific breakthroughs. In the past few years, the overall cost of a complete genome analysis has become so low that it's being used in clinical practice. Outside hospital walls the commercial pressure is rising to use genome analysis to screen healthy people for specific qualities and to see if they might develop certain hereditary conditions. A genome analysis also provides information about things that have nothing to do with medical information.

Our thanks go to the many colleagues with whom we discussed the many ethical and societal challenges related to analyzing the human genome. At the University of Leuven, various discussions took place within the confines of the Metaforum: an interdisciplinary working group of researchers and other stakeholders. At the same time we summarize here the discussions that take place at numerous national and international meetings of geneticists and ethicists. Thanks also go to Nicolas De Smidt, Leigh Jackson and Kate Elliott for the translation and editorial work.

Moreover, this book was originally established through collaboration between the University of Leuven and the International Cartoon Festival in Knokke-Heist (Belgium). More than 100,000 people visited a unique exhibition on human genetics and cartoons during the summer of 2014. We want to thank Christophe Floré, curator of the International Cartoon Festival of Knokke-Heist, who introduced us to the delightful and crazy world of cartoons and cartoonists. Some of the cartoons from this book were also exhibited during the Cartoon Festival.

And last but not least many thanks go to the Ballon Media publishing house that supported the creation of the original Dutch version of the book and to Leuven University Press that elaborated a completely new English version of the book. Their initiatives brought together scientists and cartoonists, but also wanted to make it clear that the work around human genetics, that takes place in academic centers, touches everyone.

Pascal Borry & Gert Matthijs
(KU Leuven - University of Leuven, Belgium)

1
The Genome: The Cookbook of Life

Genes are the recipes for the building blocks of life. All of an individual's genetic information is listed in a genome.

When most people hear the word "gene" they usually think about an illness, a gene malfunction or some obscure, intriguing attribute. If you think about the fact that most of our 22,000 genes are normal and are in fact very hard workers, this is rather a shame: the majority of our genes are okay and make sure that we are here and that our bodies function properly in a challenging environment.

The Complete Genetic Code in 46 Chromosomes

An individual's genetic code can be found all over their entire body. The nucleus of each of the one hundred trillion (a 1 followed by 14 zeros) human cells holds DNA stored in the form of 46 chromosomes. However, there are a few exceptions, such as red blood cells – which do not have a nucleus and therefore are not able to hold DNA – and sperm and egg cells – which are each meant to deliver half of that number to the next generation.

> **"The fact that a person has 46 chromosomes is a mere coincidence and has no significant importance.**

Because sperm and egg cells hold only half of an individual's complete genetic information, they make sure that an embryo, the result of the union between an egg cell and a sperm cell, has a normal number of chromosomes and can start developing into a normal being.

Painter

Theophilus Painter was born in 1889 in Virginia. He studied chemistry and zoology and wrote his thesis on the development of sperm cells in arachnids. He worked in Germany for a year with the famous scientist Theodor Boveri, who studied how chromosomes were inherited from a cell by its daughter cells. Painter became a professor in Austin, Texas and specialized in chromosomal research. He did his research mainly on possums and fruit flies, but also attempted to count the number of chromosomes in humans.

Chromosomes – which kind of look like little sausages in our imagination – are visible only when cells are dividing. In resting cells the DNA is like a tiny ball of wool stored in the nucleus and it is impossible to recognize individual chromosomes. It is only when a cell gets ready to divide that the DNA is rolled up and the chromosomes become visible with the help of a microscope. Painter picked sperm tubes, which have a lot of dividing cells. Since living testes are hard to come by, Painter collected his research material in the Austin State Hospital, where people who were committed to undergo the mandatory procedure of castration for 'therapeutic' reasons. After the 'donation', the testis had to be cut into tiny pieces to prepare the sperm tube for a microscopic examination. To cut a testicle swiftly, Painter designed his own cutting tool made out of small razors. When he then tried to count the chromosomes per cell under his microscope back in 1923, Painter miscounted the number of human chromosomes; an error was made that was not corrected until 30 years later.

The fact that a person has 46 chromosomes is a mere coincidence and has no significant importance. Chimpanzees, gorillas and orangutans have 48. For a long time, scientists believed humans had 48 as well because of one scientist's small mistake: back in 1923 Theophilus Painter miscounted the number of human chromosomes under his microscope. It wasn't until 1956 that Joe Hin Tjio and Albert Levan would correct this error and set the record straight.

If you know that a brown bear has 74 chromosomes, a donkey has 63, the same as a giraffe, an elephant has 56 and a cat 38, a mosquito has 6 chromosomes, a fruit fly has 8 and a carp has 108, you know that the number of chromosomes is not at all determinate of our place in the animal kingdom.

Technically, we should be saying '23 pairs of chromosomes' because all of our genetic information comes in doubles: we inherit a set of 23 chromosomes from our father and the other set from our mother. Thus 46 in total.

Chromosomes as the Carriers of Genetic Information

In our human chromosomes there are approximately 22,000 genes. In the 1980s and 1990s, based on predictions about our cells' complexity and the number of different proteins needed to create a human being, scientists estimated that there were around 100,000 genes. Since then, scientists have been able to map out the entire human genome and we now know that the exact number of genes is significantly less. One of the reasons is probably that the information in genes can be read in different ways and that one gene could possibly generate multiple proteins (more on that in Chapter 3). Genes are not placed on chromosomes in a specific order; they tend to be crisscrossed and ones that seem to belong together, i.e. genes that code for similar proteins, are usually not even close to one another. They might as well be lying on different chromosomes. There is no logic either when it comes to the order, nor is it important to know if a certain gene is on chromosome 13 or 17, except to people trying to get a complete and accurate overview. What is relevant is the difference between chromosomes when it comes to length and number of genes. Chromosome 1 is the largest and has approximately 2,000 genes. On Chromosome 21 there are fewer than 250 genes. The X chromosome has well over 1,000 genes, yet the Y chromosome has only around 50.

Each one of us inherits two complete sets of chromosomes – and because of that, two complete sets of genes as well – with one coming from our father and another from our mother. Humans are diploid, just like other mammals and just like the majority of all living creatures. We possess two copies of every gene. One could say that we have a double genome. That is the consequence and the advantage of sexual reproduction: we have a backup copy of every gene and in every generation there can be an exchange between the genetic materials of the maternal and paternal genome, which makes it possible for there to be a lot of difference between siblings even though they share the same parents. This is not the case with bacteria, which have one small, circular chromosome: if you come across a pneumococcus contagion, then all the germs that you cough up are genetically identical. That is at least until one of them mutates and becomes resistant to the antibiotics that you are using and takes action to survive.

> **“It is astonishing to see nature create so much diversity on the level of the organism based on small differences in the genetic recipe book.**

Recipes on How to Prepare a Human or Mankind

Genes are like recipes in a cookbook: they are a summary of instructions that need to be followed with surgical precision if you want the end result to be satisfactory. Every genetic recipe has the instructions for creating a protein or, more schematically, of a physical quality in the form of a protein. The difference between genes on chromosomes when compared to recipes in a cookbook is that there is no specific order for the instructions. You cannot categorize genes like you would classify lists of appetizers or desserts, or dishes with meat or fish in a cookbook. But aside from that, it is a pretty accurate analogy: our genome is a cookbook that holds 22,000 recipes.

Genetics and Evolution

Where do all these recipes for making a human come from? Throughout our entire evolutionary path, genes have been collected, adapted, replenished. A process of millions of years was required to make us who we are: humans of the kind known as Homo sapiens. It allows us to barely distinguish ourselves from other primates, such as the gorilla, the chimpanzee and the bonobo. We are only a little bit further removed from other apes, pigs and shrews. Is it not peculiar to know that in each and every one of us, in our genome, our cells, there are tracks of the genetic information that date back to our species' evolution from unicellular organisms to multicellular organisms? However, it is what it is and sometimes it can be noticeable. Our cell structure is not that different than the ones we find in more primitive species. That same thought arises when you look at how many embryos of different animal species look alike in the earliest stage and how differences only become noticeable at later stages. On a cellular level the differences can be minimal.

The principle of genetics is clear: we have a mutual genome (more specifically the genes that we need to have in common with all the members of the same species, in our case Homo sapiens). But at the same time our genome can be very personal, especially if we look at the level of the individual letters. On top of it all we have a duplicate of every genome, which makes this matter even more complicated, but also a lot more interesting.

The individual cookbook for all people is pretty much identical, and that is how it should be since they have the same goal: creating a human being. The only difference is that we like to differentiate between the recipe of mankind and the individual recipe in all of us.

The recipe book of mankind is unique even though the individual recipes or genes resemble those of an ape, a dog or a mouse, and in some cases we will share similar instructions with flies or even a banana. This is because for many proteins the function has stayed the same or barely changed over the course of the millions of years of evolution. Our cells grow and breathe just like the cells you will find in yeast used to make beer, and the proteins that play a role in the conversion from sugar to energy could be exchanged amongst each other. To put it in cooking lingo: the ingredients and amounts in a recipe can be nearly identical. Of course there also need to be essential differences otherwise a human being would not be human. That is why we talk about the human genome, the ape genome, the elephant genome, etc.

> **“As a species, all humans have the genome in common. It is the tiny differences in individual genomes that ensure that every human being is genetically unique.**

Globally speaking, the gene composition in humans and other primates is a 99% match. The quest to figure out the origin of mankind is one that tracks genes that started distinguishing themselves through the course of evolution and differentiated the human genome from organisms predating humans (see Chapter 15).

There are also differences amongst humans, and then we get to the part about the individual genome of a human being. This is important: As a species all humans have the genome in common; tiny differences in individual genomes make sure that every human being is genetically unique. Those differences tend to be hidden in the fine print. Your mother's recipe for bread pudding is just a little bit different from the one your grandmother uses. As long as we are just talking about a little more or a little less sugar, that extra scoop of butter, there are no major problems. One might prefer brown sugar over white. No problem, all recipes are good and will lead the way for you to make bread pudding. This is the same when it comes to genes: there are small variations in the text that make sure there is a different effect (see Chapter 3). The sum of all those differences enables us to be different and unique. The genetic plan of every individual, being the individual genome, is unique. The only exception to the rule is the case of identical twins.

The Genome in Single Digits

A genome consists of tiny chemical building blocks, connected like pearls on a necklace. The genome is the sum of all chromosomes, chemical data that can be read like a book: we do not really care about the paper it is printed on; we are more interested in the text. Look at the tiny chemical building blocks as if they were letters. The genetic alphabet is made up of 4 letters: A, C, G and T. With these letters we differentiate between four different chemical molecules, 'bases' or 'nucleotides'. Often it will be called a 'base pair' because for every base on one of the DNA strains in the double helix there is one corresponding base on the other strain.

An entire human genome is made out of three billion base pairs. It is a long chain of As, Cs, Gs and Ts in all possible combinations. Since people are diploid and so have a double of every chromosome, every nucleus of every cell has six billion letters. DNA is sub-microscopically small, but if you were to take the DNA of all the chromosomes in a single cell and roll it out, it would leave you with a tiny cord that is approximately 1.5 meters (or almost 5 feet) long. How much would the DNA of each cell weigh you ask? Around 6 picograms, which corresponds to 6 thousandths of a billionth of a gram (a gram = 0,035 ounces).

2
Reading Genomes

To understand what genes do, you have to be able to read the genome. This is possible today thanks to substantial scientific advancement in a relatively short period of time.

Already at the beginning of the 20th century, it was clear that nucleic acids – chemical molecules that are built out of the bases A, C, G, T – played a role in a cell inheriting genetic information and passing it on from one generation to the next. But how this happened was not very clear. In 1953 James Watson and Francis Crick, two Cambridge (United Kingdom) researchers, came up with a model. James Watson – his friends called him Jim – was a young American biologist and Francis Crick was a British physicist who was 12 years older. Their model was the double helix in which two strands of DNA are interwoven in opposite directions a bit like a long, turning ladder. The idea of this double helix took shape after they saw the results of an x-ray image of nucleic acids that had been produced by Rosalind Franklin in the London-based laboratories of Maurice Wilkins.

“The double helix is a trademark of living organisms.

Watson and Crick Described the Double Helix in 1953

Watson and Crick (in that order) published their findings on just one page (and some lines on the next) in *Nature* on April 25th, 1953. It ended up being a milestone in the history of science.

One Double Helix for Every Chromosome

The two strands of the double helix are not the two chromosomes we inherit from both of our parents. In fact, that is a small misunderstanding. The double helix is the molecular shape that the DNA takes and double-stranded DNA is further folded until it becomes a chromosome.

The double helix of DNA's task is to store genetic information so that it can be passed to the next generation without error. The chromosomes consist of DNA and are practically and conceptually considered to be the biological structures that carry genes. The genes lie on the chromosomes like songs on a music cassette: the information is imbedded in the DNA molecule just as the music is imbedded in the cassette (the latter happens under the formation of electromagnetic signals), and the chromosome is a linear piece of DNA just like the cassette contains one long thread of plastic that carries the music.

In order to have doubles of all that genetic data, you inherit one set of chromosomes from each of your parents. Every chromosome appears twice and contains all the genes, along with the genetic variation you inherited from your parents.

The double helix is made out of two DNA strands that are antiparallel: they run in opposite directions. On the opposite side of every A base on one strand there is a T on the other, and on the opposite side of every G base on one strand there is a C on the other (and vice versa). These form 'base pairs' in the double helix. The genetic data that is saved in the order in which the As, Gs, Cs and Ts appear on one strand can be found on the other strand as a mirror image. This is why genetic data can be copied perfectly: if a double strand is taken apart, every strand on its own can be used as a mould to remake the other mirrored strand by putting each A, G, C or T in the existing strand in front of a T, C, G or A in the new strand. This is how two perfectly identical 'daughter strands' are created.

The double helix is a trademark of all living organisms, not just people or mammals: insects, plants, and even the simplest bacteria and some viruses have them. However, the genome of a bacterium or virus will be a lot less complex than that of a higher life form.

Reading the Genetic Code with the Sanger Method

Which techniques enable us to read the composition of a piece of DNA and, when extended, an entire genome? One technique caused a breakthrough and is still used to this day to examine small samples of DNA or individual genes. This method was developed by Frederick Sanger in the beginning of the 1970s and carries his name: Sanger-sequencing. It is used to determine the order of the bases, the DNA building blocks.

Sanger had the idea of interweaving the building blocks of DNA – the As, Gs, Cs or Ts – in a test tube by using enzymes that nature uses to do the job. These enzymes were the DNA 'polymerases' that are capable of making a complementary strand of DNA based on the sequence of one strand. In nature, DNA polymerases copy the DNA during every cell division so that the entire genome can be passed on to the daughter cells.

Sanger's ingenious idea was to add a small concentration of chemically modified As, Gs, Cs or Ts to the mixture in the test tube. The natural DNA polymerase did not know what to do with this abnormal building block, so every time one of these synthetic bases –see them as the broken pearl in a pearl necklace was added, the reproduction of the piece of DNA would come to a standstill. If you make sure this process occurs randomly, you end up with a mixture of different little pieces of DNA that have all been copied from the same 'mould', but each time the synthesis stopped at a different position, ending with a modified A, C, G or T.

His second light bulb moment was adding a traceable marker – kind of a special bead that you can recognize as the start of the fragment you want to read. Originally this bead was a radioactively labeled group linked to the DNA; nowadays they use a fluorescent marker, which can be implemented in different ways. If you separate this mixture of fragments of different length, ending respectively in A, C, G or T in the next step at the laboratory, you get a kind of ladder that is very easy to read: ACGTTCGATCAATCCT, etc. And boom: you can start reading DNA.

Sanger ended up winning two Nobel prizes: in 1980 for inventing this method of reading DNA and earlier, in 1958, for coming up with a similar method to unravel the composition of protein, more specifically insulin. Later in his career, he led the well-known Wellcome Trust Sanger Institute in Hinxton, near Cambridge, and it was also named after him. Still today it is the largest genetic research facility in Europe where they test the latest DNA techniques. Sanger passed away in 2013.

> **“You cannot just summarize the complexity of a genome onto a simple piece of paper.**

Next Step: Reading Genomes

Sanger's technique was automated by many of his successors, including Leroy (Lee) Hood at the California Institute of Technology (Caltech). This scientist, inventor, visionary and entrepreneur developed the first automated DNA sequencer. Sanger's method had a couple of limitations. It could be used only for DNA fragments of a couple hundred bases. The usage of the radioactive markers made it very exhaustive. To top it off, the letters had to be read manually. Imagine typing 3 billion letters of a genome into a computer file (good luck finding an assistant willing to do the job). Hood replaced the radioactivity by fluorescence and refined the procedure. He also made sure that the fluorescent signals were read automatically by the machine, and translated into a DNA sequence using software. There were other machines on the market, but none could match the success of Hood's device.

Meanwhile, it was 1986 and the idea of mapping the entire human genome took shape internationally. Lee Hood's appliances came just in time. Applied Biosystems, a company from California, brought them onto the market and produced thousands of machines fairly quickly. These machines became the workhorses of the Human Genome Project, but also of every genetic laboratory and every scientific research institute. Whoever wanted to read DNA – or sequence DNA to use the scientific lingo – needed this device.

International Human Genome Project

The preparations for the Human Genome Project started in the 1980s. Supported by numerous research teams from different countries, the goal of this gigantic project was to map all the genes of a human being – a genome. In 2000 the first version of the genome was presented to the world's press at the White House. The reference genome and its updated versions of 2001 and 2003 were the result of years of hard work and a 3 billion USD investment (in the US and abroad). The result of the Human Genome Project was not the sequencing of a single individual. The first genome sequence that was published was a melting pot of the genetic data of a handful of individuals whose DNA was used to read the sequence. *The* genome that was described thanks to the Human Genome Project would serve as a reference genome that researchers would use to find their way in the 3 billion bases of an individual human genome. Only later did they sequence individual genomes. Eventually, Craig Venter's genome was the only one that was entirely sequenced using Sanger's sequencing method at an 'industrial' level.

A Neck-and-Neck Race to the Human Genome

Preparations for the mapping of the human genome were not faultless. The National Human Genome Research Institute (NHGRI), led by Francis Collins, was a part of the international consortium that set out to unravel the human genome. The whole project would be financed with public money. Craig Venter was a driven genome scientist who also worked for the National Institutes of Health (NIH) at first, but soon became frustrated because the project moved too slowly in his opinion. He decided to leave the Institute and map the human genome on his own. Well, not completely on his own: he managed to get funding from the biotech and pharmaceutical industry and started The Institute for Genomic Research (TIGR). He hired a small army of researchers and quickly bought over 300 sequencers from Applied Biosystems. As a sort of pre-workout warm-up, his team mapped the genome of different bacteria and the fruit fly.

The unraveling of the human genome became a feisty neck-and-neck race between Venter's TIGR and Collins' NHGRI. Towards the end of the race, they reconciled and joined forces to complete the project: the rivaling institutes, one financed by private capital, the other financed by the government, ended up presenting the first version of the human genome together. It went more quickly and they used a smaller budget than they had originally planned on.

Since he had all the necessary equipment at his disposal, Craig Venter continued to eventually sequence his own genome. He became the first human being whose genome was completely sequenced. It cost him around 70 million dollars. Later, when he became less active as a human genome researcher and he gave in to his other passion of sailing, he could not resist the urge to collect thousands of bacteria and other microorganisms during his sailing trips and sequence their genomes.

Massively Parallel Sequencing: How Do You Read an Individual Genome?

The original method used for mapping a genome was expensive and time-consuming. A new advance made it possible to take a completely different approach to sequence an individual's genome. The solution was to miniaturize the process and analyze hundreds of thousands to millions of short pieces of DNA simultaneously. This technique is called 'massively parallel sequencing' and uses many physiochemical tricks and high-end technological equipment.

To do so requires a lot of IT and bioinformatics. It is not just about reading the DNA order of the individual, but also all of the variations within that genome, i.e. the DNA inherited from the mother and the father. To top it off, this data needs to be interpreted. This requires not just sequencing technology, but special and complex algorithms, supercomputers and (international) databases to compare the DNA sequence and the possible variations.

This new technology to enable massively parallel sequencing soon became known as 'next-generation sequencing'. The first commercial device was created by 454, a small technology company that would later sell its invention to the pharmaceutical and diagnostic giant Roche. To demonstrate that it was possible to sequence a genome with this device, the company came out with a major stunt in 2007: the DNA of Jim Watson – who discovered the double helix back in 1953 – was mapped completely and published. This whole operation reportedly cost one million USD.

Many genetic institutes bought such a device as soon as it was released on to the market. In a clinical context, it was mainly used for analyzing BRCA1 and BRCA 2 genes (see Chapter 5), and in a research context to sequence bacterial genomes. This had to do with its capacity. This output of the first generation of "massively parallel sequencing machines" was impressive, but mapping out an

Massively Parallel Sequencing

The first step in the process of massively parallel sequencing is to cut the DNA into short fragments between approximately 50 to 500 bases. This is relatively simple: when a solution containing the DNA is mixed at a very fast pace, the double helix will break off into little pieces. The equipment that is currently on the market sticks a little piece of synthetic DNA to each of these fragments: this is necessary to be able to 'fish' them out of the solution and isolate them. Indeed, the individual fragments find their way to a glass plate etched with microscopically small holes or loaded with miniscule 'clusters'. Then their sequence is read chemically using methods that are not very far removed from the Sanger method, with the difference that everything happens on a microscopic scale, which means that the fluorescent signals that need to be recorded are extremely weak. There is a lot of nanotechnology involved with this process, with the controls of the chemical reactions as well as with the detection of the signals. Small video cameras or other sensors will register the changes that are caused by these reactions. The trick is that the signals have to be measurably different when an A, C, G or T needs to be detected. This works amazingly well with these machines, and loads of measurements are collected. Software will then transfer those miniscule differences into readable information, which causes the appliance to produce millions of pieces of DNA sequence. This is the raw data. The analysis and interpretation begin when these pieces of DNA are put next to the reference sequence and can be compared with it. Interestingly, these techniques were developed after the Human Genome Project ended. The first massively parallel sequencing device was released to the market in 2005.

entire human genome at a cost of one million USD a piece was just too expensive to use in clinical practice. Thus, by the end of 2013, most of these machines were no longer used – this model's time had passed. Not that the machines were worn down or the quality of the analysis was subpar, but there were newer machines available that had a higher capacity while the cost of sequencing a DNA fragment was 10 times less that on the Roche machines.

Sequencing Machines Keep Getting Faster and More Powerful

Applied Biosystems (Life Technologies) and Illumina released devices that could sequence a genome more easily. They work based on the same basic idea: they can simultaneously execute hundreds of thousands to millions of DNA sequence reactions. The difference lies in the physiochemical approach. Illumina is the market leader in this field at the moment and expands the capacity of its devices approximately every 6 months. Life Technologies equally releases new versions of its devices every year or so. Roche does not have a good follow up for its first system, but this can change rapidly: there are different companies and research institutes that are developing (even) new(er) methods. As soon as it becomes clear that a technique has a good commercial chance, a big player will pick it up and funding will become available. It is hard to predict on which technology the *next-next-generation* devices will be based and which producer will offer it.

In the meantime, the race to read DNA molecules as they appear in their cells, without any enrichment of materials, has started. The term for this is *single molecule sequencing*. There are already devices on the market that can read the composition of individual pieces of DNA directly (and some have already disappeared from the market because they did not meet expectations). At the moment Pacific Biosciences' device is leading the race. It is not used for clinical purposes at this time, but they are working hard to make it possible. It is not certain that this will work before yet another technology replaces the current one. In other words, research is going extremely fast and staying with the game is not easy, especially when a new machine costs millions of dollars. The next big thing is when the manufacturers make the devices themselves smaller and cheaper, so that individual research teams and clinical laboratories can again afford them. The alternative is that only a handful of research institutes or private companies buy these super expensive machines with enormous capacity, and offer sequencing as a service so that the price of sequencing in itself is reduced. The downside of the latter is that all the research and diagnostic activities in the field of genetics become increasingly centralised.

China Steps In

The result of the technological race and the interest of investors started playing a role in the commercial and strategic developments in the field of genome sequencing. BGI (Beijing Genomics Institute), a Chinese research institute and company, shook up the world of DNA sequencing when it installed approximately 140 Illumina 'massively parallel sequencing' devices and offered *sequencing services* worldwide on a commercial basis. When BGI decided to bring some spare devices from China to Europe in 2012 to offer their services locally, an agreement with Illumina went down the drain. Illumina took this gesture as a form of competition for the European market, and this using its own devices.

In 2012 Complete Genomics, a company that offered a sequencing service based on a technology that was developed in-house, was bought by BGI. Illumina fought quite a bit for the Complete Genomics takeover, but eventually had to give up. Nevertheless, as long as they have to deliver appliances for their devices installed at BGI, Illumina has the upper hand.

Early in 2014, Illumina announced that it had broken the '1,000 dollar per genome' barrier. This makes reading the entire DNA sequence of an individual relatively cheap. But we have to point out two things. First of all, the company admits that the write-off of the device is not included in the price, and this would cost around 10 million euro (or, better put, you would need ten machines worth a million dollars to do the job). And second, 1,000 dollars is the cost of reading an entire genome sequence. The costs linked to the interpretation of all that data are a lot higher, especially if the interpretation is meant to identify a genetic defect and to genetically counsel patients and their families. The generation of the sequence data is mainly executed by bioinformaticians, whereas the interpretation needs to be done by specialists in the field of molecular genetics or experts with a clinical background, or a combination of both.

For small countries and smaller research institutes, genome research will become a challenge in the future. There might be smaller devices available in the near future that are affordable for smaller labs and that can do relatively small tests, for example for hereditary breast cancer. Alternatively, big institutes or governments may invest in very large devices and install them in national or international laboratories that will have a kind of monopoly on big studies and attract an important part of the sequencing for clinical purposes. The latter will be a thorn in the flesh of many researchers, but in the end economic factors might prevail.

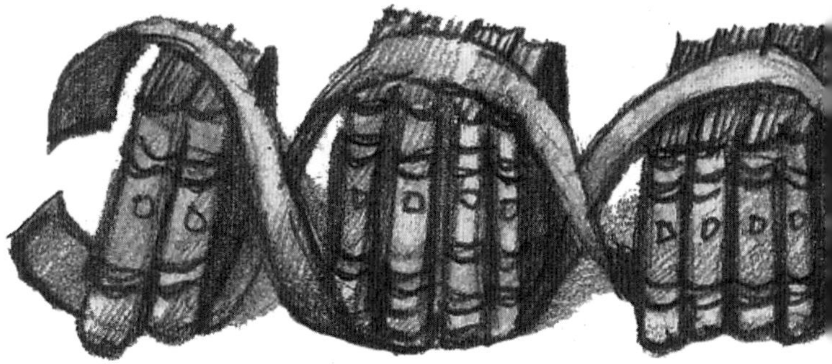

Crick Together with Watson, Watson vs. Venter, Venter alongside Collins

In 1962, Jim Watson, together with Francis Crick and Maurice Wilkins, received the Nobel Prize for discovering the double helix. Rosalind Franklin did not share in the victory because she had already died and the Nobel Prize cannot be granted post mortem. This probably worked in favor of the other 3 laureates since the Nobel Prize committee has never awarded one prize to four scientists.

Watson and Crick both remained active in scientific research. Crick especially made some fantastic discoveries: he was part of the unraveling of the genetic code, that is the 'alphabet' that is used to convert the information embedded in a DNA fragment into the production of a protein (see Chapter 3). Francis Crick died in 2004.

Watson returned to the United States and became the head of the Cold Spring Harbor Laboratories, a large research institute on Long Island. From 1988 until 1992 he was director of the Human Genome Project, the international academic consortium that mapped out the entire human genome for the first time. Craig Venter, another important American genome researcher, who left the National Institute of Health and created a research institute with private funding, does not cut Watson any slack in his book My Life Decoded. He accuses him of a lack of vision and his backstabbing approach to the industrial competition against the academic consortium that was funded by the U.S. Congress. Venter probably tells a one-sided story, but it remains a fact that Watson in 1992 was shown the door as a director of the Human Genome Project.

Watson's disciple, Francis Collins, took the project back up to speed. The

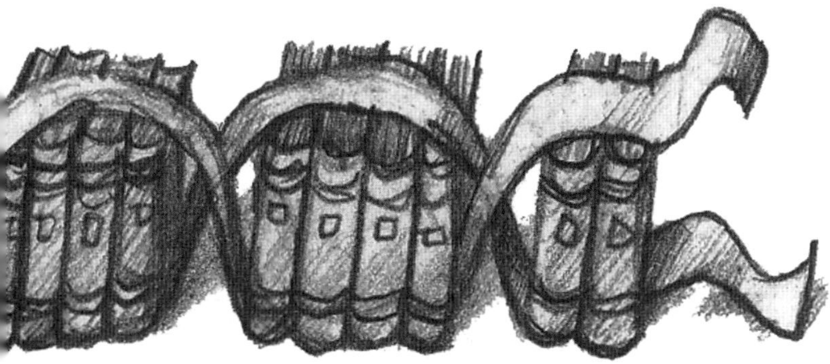

enterprise was fueled by the tension that emerged when Venter invested in hundreds of machines that could read millions of DNA letters daily. Venter used a different technique and a supercomputer – it was at that time the second largest computer in the world and the largest computer of its caliber in private hands – to puzzle the pieces of hundreds of millions of fragments of some hundreds of letters into an entire genome correctly. His team –and his computer – got the job done in a couple of months.

This put an enormous amount of pressure on the international consortium that had decided beforehand to first map out the genome – and built a kind of a road map – and then sequence it in smaller parts. Eventually, they both reached the finishing line at the same time more or less, in June 2000, Collins and Venter were invited to the White House to present the first version of the complete genome to President Clinton. A 'forced tie', according to Venter, who has few good words for Collins in his book.

Watson and Collins both wrote books about the International Human Genome Project, with a different perspective than Venter of course. The truth probably lies somewhere in the middle. In any case, the unraveling of the human genome showed respectable entrepreneurship and cost billions. Venter, Watson and Collins all take merit from the success, together with thousands of researchers, technical crew and bioinformaticians who worked on this project worldwide. Why Collins and Venter never received a Nobel Prize remains an open question. Maybe because of the political bickering, or maybe because of the fact that a lot of other important scientists also made essential contributions.

3
A Gene, a Genome, and Then What?

The genes in the genome are the blueprints of the proteins a (human) body contains. Because genes – and the proteins they encode – vary, we all look different.

After the double helix had been described in 1953, it took researchers a while to gain insight into the way the information that is contained in the sequence of As, Gs, Cs and Ts in the DNA is converted into an instruction to produce proteins.

A breakthrough was found in the 1960s. Sydney Brenner, a South African biologist (and, yet again, winner of the Nobel Prize) played a crucial role in decrypting the 'genetic code', but Francis Crick also made an important contribution. Proteins are the actual building blocks of cells and they play a role in the many biochemical processes that go on in our bodies. There are as many functions as there are proteins: they can break down nutrients and convert them to energy; they make sure we take up oxygen in our bloodstreams and that oxygen is transported; they enable our cells to interact and ensure that our joints are flexible; they are deployed in our skin for protection; they translate our brain signals into understandable instructions for our nerve cells; etc.

Are There Any Typos in the Text?

If the order of the letters in the DNA is strongly connected to the composition of the proteins that are derived from them, then every variation of a gene will have a possible effect on the function of the corresponding protein. A mutation, speaking in a chemical way, is when a base in the DNA is replaced with another base. For example, replacing a G by a T, or replacing an A by a G.

In practice we make a distinction between neutral variations that have little to no clinical effects and damaging variations that cause clinical problems. The former are called *polymorphisms*, the latter are referred to as *mutations*.

Polymorphisms are not considered to be damaging, but they are not necessarily neutral either. Instead, they can be looked at as being little typos that snuck into the DNA but that do not have clinical significance. They just warrant that we all look different!

Mutations on the other hand do have a significant and often unfortunate effect on the functioning of a protein. This is why the word *mutation* got a reputation for being bad or unwanted, which – strictly speaking, at the chemical level – is not necessarily the case. Therefore, the term "pathogenic mutation" is often used to refer to changes that have a detrimental effect. The adding of pathogenic is superfluous, but it adds clarity. We will sum up a couple of examples of pathogenic mutations in Chapter 4. Let's first have a look at what polymorphisms entail...

> **“Proteins are the actual building blocks of cells and they play a role in the many biochemical processes that go on in our bodies.**

The Code for Proteins

The four bases in DNA – A, C, G and T – can be seen as the four letters of the genetic alphabet. A cellular 'deciphering' machinery moves along the DNA strand and reads the letters in groups of three. The words in the genetic alphabet consist of three letters, so-called codons. If you can use four letters to make three-letter words, you end up with 64 possibilities.

Therefore, in DNA we can find 64 different words. In normal language you use words to build sentences, and with sentences you can write a novel, a newspaper article or an encyclopedia. In genetics the four letters are enough to write the three letter words. These words then form sentences that contain the code for proteins. These sentences are called genes. All genes combined form the genome. The specific code makes sure that the genome of a human being comes to life as a human being and not as a frog or daisy.

Proteins are made up of amino acids, and in total we know of 20 different amino acids (in humans and most other living organisms). How do you link genes to proteins? Each codon corresponds to a specific amino acid. Thus, since there are 64 different possible codons, the system can easily specify the amino acids that need to be used (actually, there is even some redundancy in the codes, but this is not important). A gene thus contains a sequence of codons that code the amino acids. All genes begin with a *start codon* and end with a *stop codon*. Each cell has this fantastic machinery at its disposal that scans DNA and reads the code, and based on the code it assembles the amino acids into proteins, like links in a chain, until the code of the gene has been completely converted and the protein is ready for use. Genes are copied one by one because they are used separately – like the recipes in a cookbook. A particular cell will use only those genes that are necessary for its function and survival. Thus, each cell uses the information contained in the genome in a different way.

Strictly speaking, there is one more step in between: the relevant piece of DNA is copied to a "messenger" before it can be translated to protein. This is because the DNA is stored in the cell nucleus while protein production happens in a different part of the cell. You can compare it to a library with important books: those books – the DNA – cannot leave the library, but you are allowed to copy a part of them to use as a plan for your building project – in this case to produce proteins.

A lot of scientists have contributed to decrypting these mechanisms. It would not be fair to limit ourselves to the contributions of Watson, Crick and Brenner.

Aside from the "neutral polymorphisms" and "pathogenic mutations", geneticists often use the term "variants of unknown significance" (VUS). This term is reserved for those cases where it is not yet clear to what extent the "typo" will affect the functioning of the protein. If you find this term (or, in some instances, the term "possibly pathogenic variant") in a clinical report, it means that it is not entirely certain that the cause of the genetic problem has been found in the patient.

The Difference between Variation and Deviation Can Be One Letter

Everyone starts out with a spectrum of variations in his or her genes. The question is whether such variations are natural changes without a real big impact, or the causes of disease. Back to the cookbook. What if the recipe for bread pudding all of a sudden changes and tells you to use 15 ounces of sugar instead of 4? Maybe it would not be that bad if you have a sweet tooth. But if the recipe told you to use salt instead of sugar, you would immediately know that it would not turn out to be normal bread pudding. That moment the normal variation changed into a defect, into a gene that loses its normal functionality, it turned into a recipe that lost its meaning.

In some cases a small genetic typo can mean a world of difference. A typical example is dwarfism. Almost all patients that suffer from achondroplasia (clinical term for dwarfism) have a mutation on the FGFR3 gene. Nearly all patients worldwide have the same genetic error, the same different letter. FGF is a type of growth factor, and the FGFR3 is the "receptor" for this, a kind of antenna. In normal conditions, the receptor makes sure that the signal of the growth factor is clearly received by the cells of the long bones. However, a genetic error in the receptor disturbs this reception and causes a growth disorder.

You can compare genetic errors with a recipe that misses out an ingredient or provides wrong guidance. This results in the recipe, the defect gene, not being able to produce a normal protein. This can have an effect in many different ways. Either there is no longer a protein and you lose a function, or the deviant protein gets a new function because of the typo, one that it normally would not have.

Both scenarios are severe and lead to deviations that could mean the difference between life and death. Some deviations are so drastic that the fertilized egg cells or the embryo die during pregnancy. If the essential life functions are not supported because the new genome contains too many errors, this will lead to a spontaneous miscarriage.

When a defect hits one gene, geneticists call it a "monogenic defect" or "monogenic trait".

"Normal" Traits Can Be Purely Genetic, and *Vice Versa* – Genetic Variation Can Be Really Normal

Tongue rolling for example is a monogenic trait. Either you can roll your tongue or you cannot. There is only one gene that is responsible for this and it comes in two variants: the variant that enables you to do so and the variant that makes sure that you cannot. The gene is unknown, and to this day no one has wasted research funding to study the cause of this trivial trait. So today we cannot say if the gene functions properly in the first case or if it does not function in the other, or why the gene has been conserved in two versions throughout evolution. What is the normal function of the tongue roll gene? Does it have anything to do with the tongue muscle, the nerves in the tongue, a brain function? We do not know but it is clearly a monogenic trait.

Besides, the effect of the tongue roll gene is "dominant". If a child can roll his or her tongue, then the child has at least one parent who can do so as well.

More Variety than Expected

If the code of all genes is compared amongst people, it shows that the variation is approximately 0.1%. On average you will find one difference for every 1,000 base pairs. It means that the letters A, G, C or T have been replaced and cause variations between individual people. These positions, on which differences are possible, are called SNVs (single nucleotide variants). People can be recognized and identified based on their SNVs. This makes genetic variation useful for forensic research. But if the SNV hits an essential base pair in a gene, it leads to a genetic deviation (or mutation, as was discussed before).

A couple of years ago a different form of genetic variation was discovered: one that consisted of doubling or even multiplying entire pieces of our genetic material, comparable to entire pages of a cookbook that are copied once or even many times. These are the so-called "copy number variants" (CNVs). These variants are not necessarily pathogenic: everything depends on the function of the genes – and their proteins – that are contained in the CNVs. However, some CNVs can influence health and disease. In contrast, sometimes whole fragments of the cookbook can be missing. In this case, the CNVs correspond to so-called "deletions". Again, deletions can either be neutral or be implicated in genetic variation and disease.

There are also abnormal genomes that are missing an entire chromosome or have one too many. In that case we are dealing with numerical deviations in the number of chromosomes. A defect like this is almost always associated with severe disorders. Some of these numerical deviations result in early death, most often early in embryogenesis (more on that in Chapter 4).

Dominant versus Recessive

For a good understanding of the terms *dominant* and *recessive* we have to go back to genetics and cell biology. Genes code for proteins and we have inherited two copies of each gene: one from our father and one from our mother. Normally, cells use both genes to produce the corresponding protein. But sometimes mutations cause one gene to take the upper hand. Or, better put, the protein of one gene has a dominant function compared to the protein of the other gene because of the mutation.

You can compare this to the two wheels of a bicycle. At first glance, both tires look alike because they have been created through the same process. If you run a flat tire in the front, you have a problem. It does not matter if your rear wheel is still fine, you cannot keep on riding. The effect of the flat tire is dominant: it does not help that you have a perfect back up of the gene; the function of the altered protein is so disrupted that the normal protein can no longer fulfill its purpose.

In contrast, in the case of a recessive inheritance pattern it can, because the malfunction of the altered gene or protein does not influence the functioning of the normal protein. Imagine that every car had two batteries, but that the energy of one battery was enough to start the car and make it fit to drive. You would not notice it if one of the batteries was malfunctioning even though it was broken from the start. It is only when the second battery gives up and the car does not work anymore that you would notice that the one of the original batteries was also malfunctioning. If the car manufacturer installs two broken batteries, the car will never be able to drive. This can be compared to a hereditary recessive defect: whoever inherits a defect in the same gene from both of his or her parents will have a trait (or a disease) that was unnoticeable in the parents.

Eye color is not a strictly monogenic trait, but certainly blue is "recessive". This means that someone with blue eyes has to inherit a gene variant that codes for the blue variant from both parents. But this does not mean that both parents need to have blue eyes themselves. It is enough that they both carry one blue variant: if they combine it with the variant that codes for brown eyes, they would have brown eyes, because the brown will take the upper hand:

brown is dominant. If they both incidentally pass the blue variant at the same time on to their child, the child will have blue eyes. The child's eye color will thus be completely different from mom's and dad's. People would think, "This child has radiant blue eyes but the parents do not. Is he really the father?" Be careful: non-paternity in this case cannot be concluded from the simple fact of the child having blue eyes. So you had better not jump to conclusions, and make genetic mistakes! Remember that we have two copies of every gene and that some traits will be masked if they are recessive.

Skin Color

Modern man comes from ancestors who lived in Africa, some of whom moved from Africa to Europe and Asia, between 100,000 to 50,000 years ago. It is suspected that the original *Homo sapiens* had a dark skin color, just like most modern-day Africans.

A darker skin color protects against the harmful effects of sunlight – damage to DNA, damage to some essential substances such as folic acid – and probably has an effect on immunity. The biological choice for dark skin in areas with a lot of sunlight is clear.

So then why are most Europeans white? A possible explanation is that they have adapted to the limited availability of sunlight in areas further removed from the Equator, because sunlight on the skin also has some useful effects. Vitamin D, for example, is produced when the skin is exposed to UV light. In areas with less sunlight vitamin D production will be less efficient in someone with a darker skin color. Observation shows that people with a darker skin color in areas with less sunlight often suffer from rickets, a deficiency in the bones that is caused by a lack of vitamin D (it was very present in children of working class parents during the Industrial Revolution because they did not get enough sunlight in the workers' neighborhoods in big, smoggy cities). When *Homo sapiens* migrated north thousands of years ago, individuals with lighter skin color probably had an advantage and so natural selection happened. This was enabled by the

Variation within Normal Boundaries

In most cases, genetic differences between people are subtle and are considered to be variations within normal boundaries. This variation makes sure that we are all different. But it also contributes to our sensitivity (or susceptibility) to environmental factors, sickness and the ageing process (see Chapter 4). Variation in itself is not a bad thing. Neither are mutations. Thanks to DNA being prone to mutation, living creatures were able to evolve into complex organisms and diverse species throughout evolution. On the level of an individual person, a mutation can be devastating, especially if it causes severe illness or inborn deficiencies. But another mutation can give an individual an advantage. If that advantage leads to better survival and better procreation, the species as a whole will benefit from it. Mutations and natural selection go hand in hand.

limited amount of sunlight having less harmful effects. Thus, the advantage of a biological selection for darker skin color became less important.

Skin color, hair color and the color of the eyes are noticeable trademarks of appearance. But they are determined by only a handful of genes. Comparing people based on the sole premise of their skin color is senseless. Nothing in the human genome justifies racial discrimination.

This does not take away the fact that some groups of the population have specific genetic traits. For example, many Asians have a deficiency in one of the genes that plays a part in the breaking down of alcohol. The first step in the process will happen too fast. As the other enzymes, downstream in the process, cannot keep up, the fast processing will lead to high concentrations of the toxic acetaldehyde, a substance that has adverse effects. It leads to the so-called Asian flush syndrome. Thirty to fifty percent of Chinese and Japanese people carry this deficient gene while this deficiency can hardly be found in white Europeans or Africans, and is extremely rare in Mexicans. It is very unpleasant but, from a genomic point of view, it is just a detail.

Thanks to DNA having the potential to mutate, living creatures were able to evolve into complex and diverse organisms and species throughout millions of years of evolution.

The Quest for the Role of "Junk DNA"

The human genome consists of more than just genes. Even better: only 1% of the letters A, C, G and T appear in the genes and the rest fill in the blanks in between. We are talking about huge blanks here and, in all honesty, we do not really know what this extra mass of DNA does. It is like adding irrelevant newspaper articles, including advertisements, classifieds and obituaries pasted into a cookbook or encyclopedia.

There are most certainly fragments amongst the junk DNA that are key to survival and the cell's function. But in all honesty, not very much of how this works is understood or known.

The above-mentioned analogy is surprisingly accurate: the 'obituaries' in our genome are the remains of genetic information of viruses that infected our cells during evolution. They have landed into our genome, and stayed there like a pile of junk. In addition, our genomes contain fragments of genes that have become redundant, so-called pseudo genes. In other words: there is a lot of rubbish in our genomes.

For a while now, researchers have been convinced that organisms and cells would not bother copying all that junk with every cell division, just for fun. There is most certainly material amongst the junk DNA that is key to survival and the cell's function. Up to now, we have known about two or three different kinds of genetic information that is hidden between junk.

First, there are segments in the DNA that contain codes (instructions) to direct the use of specific genes. Not all genes are needed at the same time. A liver cell will use the genome in a different way than a nerve cell, while genes that are crucial during the development of the embryo and cause growth have to be turned off as soon as the organism is fully grown. If not, development will go wrong, and lead to cancer, for instance (see Chapter 5).

There are also regions in the genome that do not hold any particular coding information, but probably simply (or not so simply!) shape the composition of the chromosomes in a three-dimensional structure. Chromosomes are all stacked in the cell's nucleus, and the way that the DNA is folded in that tiny cellular compartment is probably important. Compare it to a ball of wool that you can either roll up nice and tidy or just scrape together. At first glance, the DNA in the cell's nucleus looks more like the latter, but it is obvious that there is method in the madness and that the structure is important for the functioning of the genes. How? We know very few of these details.

So then we are left with the remainder of DNA that certainly fulfills other, unknown functions. Scientists came to this conclusion because of the fact that some segments of junk DNA are almost perfectly conserved among different animal species. If evolution put so much pressure on maintaining this DNA, it has to have a function, even if we do not have a clue at the moment what it does exactly.

4
Genetic Disorders

Genetic variation is normal. But sometimes it goes wrong: variation can lead to a hereditary disorder. In such disorders, the dominant and recessive hereditary patterns play a role.

Tongue rolling is dominant, blue eyes are recessive. Other trademarks that come from monogenic and dominant genes are: black hair, long eyelashes, hair on the back of your hand, wide nostrils and freckles. Other trademarks that are most likely recessive are: short eyelashes and attached earlobes. And these are just the visible trademarks. You can imagine that there is a lot of variation in the different processes of our body. This is pretty unnoticeable unless the effect is indirectly detectable.

A somewhat odd example is the digestion of asparagus. Asparagus contains substances that hold sulfur, and when they are broken down by our bodies, our urine will have a distinct smell. This conversion does not happen the same way for everyone because of the genetic variation in the enzyme that is involved with the breakdown. This leads to a distinct smell in the urine of some people after eating asparagus.

“Not every child inherits every trait from both parents.

Asparagus also has another notable property: there is also a genetic variation in the way people detect this distinct smell of urine: some people can notice it, others cannot. The reason being that we need receptors in our noses for everything we smell. Although the scientific evidence is not complete, this effect is probably also monogenic (and dominant). Thus, since there are (at least) 2 different genes involved in digesting and smelling the sulfurous components, there are (at least) four possibilities concerning asparagus and scent: you produce this specific odor and you can detect it, you produce it but you cannot detect it, you do not produce it but you are able to detect it in other people (to find out you might have to accompany someone to the bathroom), or you do not produce it and cannot detect it (in this case you usually do not have a clue what people are talking about).

It appears that different monogenic traits can be present simultaneously but are inherited completely separately from one another. Gregor Mendel already discovered this in his peas: wrinkled or smooth, yellow or green, the choice is directed by different genes that are passed to the next generation separately from one another. This is an important principle of genetics: monogenic qualities remain independent. Or, to express it in family terms, not every child inherits every trait from every parent.

From Innocent Traits to Disorders

No one views one of the abovementioned traits as being abnormal. But sometimes life dances on the edge of the abyss. A mutation in one gene can lead to a noticeable deficiency. Then we are no longer talking about monogenic traits but rather about monogenic disorders. Achondroplasia, or dwarfism, was an example in Chapter 3, the remarkable result – as far as body length goes – of a one-letter difference in the FGFR3 gene. There are thousands of genetic disorders that are the result of such monogenic deficiencies. Theoretically the total number should be 22,000 matching the number of genes in our genome. Practically it is closer to 6,000. One reason is probably that for many genes a deficiency is not tolerable, in which case a fertilized egg cell or embryo will not be viable. Another reason is that not all hereditary disorders have been discovered. Current progress in genetics offers advantages when it comes to this, especially because of the possibility of "reading" all the genes of a patient suffering from an unexplained hereditary disorder. Thanks to genome analysis of specific patients, new genetic disorders are described on a daily basis. The count is still ongoing and 6,000 is definitely not the end.

Why Don't Hereditary Disorders Disappear on Their Own?

Most genetic disorders are rare. Evolution does not like them. Patients are often severely ill and, as a result, they cannot pass the genetic defect to the next generation. There is a strong pressure of selection against mutations with devastating effects on the individual and his or her family. Why is it that these disorders do not disappear? One of the reasons is purely genetic: in nature, new mutations, so-called "de novo" mutations, keep appearing. Such mutations are especially noticeable in severe dominant disorders. With achondroplasia (dwarfism), for example, 80% of the patients have a "de novo" mutation. This means that both parents have normal body lengths but their child shows signs of dwarf growth because of a coincidental new mutation in the FGFR3 gene. In the other 20% the mutation is passed on by one of the parents who has the same genetic condition. Something comparable can be seen in neurofibromatosis: in approximately 30% of the cases there is no family history of the syndrome. In recessive and sex-linked disorders, new mutations also occur, but they do not necessarily appear immediately (see Chapter 24). In the case of recessive disorders, there may be selective forces that give a certain advantage to those that carry the defect.

We see this in sickle-cell anemia (SCA) where the people who carry the mutation have a lower sensitivity to malaria. When it comes to cystic fibrosis (CF), the hypothesis is that carriers of the mutation had an advantage in times of cholera and typhus. Therefore, these individuals had a higher chance of survival in past centuries. However, there was no such advantage for patients that carry the defect on both genes: their life expectancies were low, and they died at an earlier age than average. Nevertheless, CF mutations were enriched in the population, thus the selective advantage of the carriers must have been significant. The advantages that came from carrying this mutation of course are not the same in today's world.

Cystic fibrosis is the most common hereditary monogenic disorder in the Western world. This illness affects one in every 3,000 children in Western Europe. In Africa this disease is less frequent. There you will find sickle-cell anemia to be the most common recessive disorder, with a ratio of 1 to every 50 people in areas where malaria can be found.

Victor McKusick and OMIM

Starting in the 1960s, an amiable clinical geneticist at Johns Hopkins Hospital in Baltimore kept a list of hereditary disorders. His name was Victor McKusick. In 1966 he published his first catalog titled *Mendelian Inheritance in Man* (MIM). He read all possible literature and publications on genetics and summarized all this information in his book. This book was published a number of times, with the 12th edition making its appearance in 1998. Since this edition, the list has been updated online (Online *Mendelian Inheritance in Man* or OMIM). The OMIM website is considered to be the authoritative reference by all geneticists. Whoever is looking for a short description of a rare hereditary disorder can visit the website: www.ncbi.nlm.nih.gov/omim

McKusick also described a number of hereditary disorders for the first time, one of which carries his name. It is a recessive disorder that he discovered in the Amish, an isolated community in Pennsylvania.

Other examples of monogenic disorders are Duchenne muscular dystrophy (DMD), a muscle condition that mainly affects boys because the genetic defect is located on the X chromosome, spinal muscular atrophy (SMA), another severe muscle disorder that is recessive, and neurofibromatosis (NF), a dominant syndrome with deviations of the skin and the nerves that can also cause a different development disorder. Phenylketonuria (PKU) is a typical example of a large group of metabolic disorders that are usually inherited recessively.

The list of monogenic, hereditary disorders is long. It is mainly about disorders that are noticeable at the time of birth or start to take shape at a young age. But also neurological disorders that develop later in life, such as Huntington disease, some types of Alzheimer's, Parkinson's, amyotrophic lateral sclerosis (ALS, also commonly referred to as Lou Gehrig's disease) or congenital heart conditions are monogenic and dominant.

Rarities with a Large Impact

The thousands of different monogenic disorders are relatively rare and the number of people affected by each one of those disorders is rather limited. However, in total, it is estimated that around 30 million people in Europe are affected by a monogenic disorder. To promote research, diagnosis and – when possible – treatment of these rare disorders, the European Commission issued a directive that obliges every European Union member state to have or develop a national plan for rare diseases. The goal is to map out national and international expertise and bundle that expertise in specialized institutes so that patients and families know where to go to get the right diagnosis and treatment for their disorders. The secondary goal of these centers is to advise the families and caregivers of the rare disease patients, and to promote the uptake of the costs of genetic and other tests, and eventually of treatment and therapies, by the health care system. Similarly, in the US, Canada and Australia, similar national initiatives have been undertaken for the promotion of knowledge on rare diseases and to improve patient care.

Skin and Hair Color Are Recessive

In October 2013 a 7-year-old blond girl was taken away from her Roma parents in Bulgaria because a neighbor suspected that the child might have been kidnapped. A similar case happened in Ireland shortly afterwards. A gypsy child with blue eyes and blonde hair... That raised suspicions. For those who understand genetics, such suspicious action comes as a shock. Blue eyes and blonde hair are recessive traits. The frequency of these genetic variations appearing in Roma is low, and very few people 'carry' the 'blonde' and 'blue' variants of the respective genes. If both parents coincidentally carry the variants and simultaneously pass on the "blue" and "blonde" gene variant, they will have a child with blue eyes and blonde hair. A child like that will definitely stand out in a group of people that predominantly have dark hair, brown eyes and a darker complexion. Compare it to a yellow tulip in a field of red ones. So yes, sometimes Roma parents do give birth to a blonde child.

Not All Traits – and Disorders – Are Monogenic

Body length, build, intelligence and other traits have a genetic component but they are not monogenic. Different genes contribute to how strong, big or intelligent someone will be. In these cases we are talking about "complex" heredity as opposed to the "simple" inheritance of monogenic traits and disorders. In addition, these complex traits are usually not purely genetic. Everyone knows that body length and body build will also be the result of eating habits and body activity, while your environment can influence your intelligence level. Therefore, complex traits are better defined as "multifactorial". They have a genetic component, but they are also influenced by outside factors. The spectrum is not limited to normal traits. Predisposition for heart disease, diabetes and dementia is also multifactorial. Even the ageing process is a multifactorial trait – or disease, if you like. As a conclusion, multifactorial diseases are not rare at all. To what extent is a multifactorial condition influenced by a genetic or environmental component? This varies and is not easily measurable. A way to measure this is by studying twins.

Complex Traits Are Not Usually Purely Genetic

Identical twins are genetically identical (because they originate from one egg cell and one sperm cell that sprouted two embryos) whereas fraternal twins (two separate conceptions after a double ovulation) are no different from normal siblings. If a certain trait or disorder is found more with identical twins than with fraternal twins, there is definitely a genetic component in play. The more the occurrence of a given trait or disorder differs between identical and fraternal twins, the bigger the contribution of the genes. An example is intelligence: the IQ of identical twins will be the same in 70% of the cases, whereas the odds are 50% percent when it comes to fraternal twins. If IQ was a trait that was purely genetic then there should be a perfect match with identical twins. Comparing adopted children to their biological and their adoptive parents is also a good indicator to see the difference between *nurture* and *nature* – education vs. hereditary traits. However, there have not been enough studies of twins and adopted children to have data on every trait or disorder.

Which environmental factors are important? This varies and is not always easy to determine. Mark Henderson, science editor for *The Times,* describes it in his book *50 Genetics Ideas You Really Need*

to Know as: "Parents can teach their children skills like playing the piano, but they will not be very influential on their musical talent." Someone with a talent for playing the piano will probably enjoy it more, which could lead them to search for more music. This makes it hard to estimate how big a role the child's environment or genes actually played in the whole thing. Aside from nurture and nature there is another key player: (bad) luck, for instance when it comes to cancer (see Chapter 5). Even infectious diseases that are not genetic according to our criteria (because they are caused by germs like bacteria and viruses, which do not distinguish between identical and fraternal twins) can have a genetic component: an individual's sensitivity to infection is determined by his or her immune system, which has different components that are genetically driven.

Can a Genetic Deficiency in Donor Sperm Make a Child Sick?

A while ago you could read the following headline in some newspapers: "Sperm donor infects children with neurofibromatosis". A geneticist finds such a misconception repulsive. Genetic disorders are not contagious! How can we rid the world of such a misunderstanding? E-mailing the newspaper does not help. So we will explain it here.

The article was about an exceptional phenomenon. Between 2007 and 2012 different children developed the hereditary disease neurofibromatosis and it was discovered that they had all been conceived with donor sperm. Research showed that all of these children had been conceived in different IVF facilities in different countries using the sperm of the same Danish donor. Further investigation showed that the sperm of this donor had been used for at least 43 different in vitro fertilizations. In total, there were 9 confirmed cases of children with this disorder. This leads us to two important questions.

First of all, why would they use the sperm if the donor had a hereditary disorder? Neurofibromatosis is a dominant condition. The answer is that the man does not show physical signs of the disorder. The mutation would have happened in his germline, these are the cells that develop the sperm cells. The donor himself did not have any physical symptoms of neurofibromatosis. So it was impossible to diagnose him with this specific disease.

Chromosomal Defects Also Play a Role

Among these stories about monogenic and multifactorial disorders, an important cause of genetic disorders should not be forgotten: chromosomal defects.

Sometimes a fertilized egg cell will miss an entire chromosome or will have one too many. The defect usually occurs when the germ cells are formed and the division of the chromosomes in their respective daughter cells does not happen correctly. It is like a whole chapter from the cookbook is missing or has been printed twice. This can have serious consequences. An example is Down syndrome where the extra chromosome 21 (trisomy 21) is at the base of the clinical symptoms. Chromosome 21 is relatively small, which makes doubling a possibility – or at least compatible with life. Other chromosomes contain more genetic information. If they are

Second question: why 9 out of 43 cases and not 50% as expected with dominant conditions? The answer is that he is *mosaic* for the mutation and that only some of his sperm cells were produced by cells that show the mutation. If he had the mutation in all of his cells, in theory half of his sperm cells would have the deficiency. This confirms that the mutation happened in a later stage of the man's (embryonal) development, one cell somewhere in the germ-line, which would end up producing sperm cells.

Bonus question: would it not be better if all donors were tested to exclude a genetic condition? The answer: they already test donors for the most frequent genetic disorders like cystic fibrosis. But it is not possible to test donors for all possible genetic defects. The new methods of massively parallel sequencing and genome analysis (see Chapter 2) could enable this nowadays, but there are different reasons to not do this at the present time. One of those reasons is that if the analysis were performed on a sperm sample, i.e. a mixture of sperm cells, mutations in individual cells would go undetected. Thus, one would have to sequence individual sperm cells. However, you cannot examine a sperm cell and then still use it for in vitro fertilization. A sperm cell is like a match: if you try to see if it works first, you cannot use it afterwards.

Anyway, the reply to the newspaper's unfortunate headline is: bacterial and viral diseases are contagious, genetic disorders are NOT.

missing or are passed on twice, the impact will be so drastic that it is often not possible for the embryo to survive. Most embryos with numerical chromosomal defects are lost long before birth. The other exceptions are trisomy 13 and 18 embryos, which will either die or continue to grow into a severely disabled child who will usually pass away shortly after birth. Numerical deviations of the gender chromosomes X and Y are viable. People with three X chromosomes (triple X syndrome) are not or hardly noticeable, just like men with an extra Y chromosome (47, XYY). Men with an extra X chromosome have Klinefelter syndrome (47, XXY), which is often discovered when they are trying to have children of their own, because the syndrome is associated with fertility problems. Women who are missing an X chromosome (45, X) have Turner syndrome. Chromosome deficiencies often present themselves as "chromosomal accidents", in other words they occur *de novo*, not to mention a few exceptions. Therefore, every chromosomal anomaly is an accident of nature. Also smaller copy number variants in the genome (see Chapter 3) that contain a couple of genes can lead to specific inborn diseases. The list of clinical syndromes – and clinical symptoms – caused

Exome Analysis with Hereditary Disorders: Pretty Nice, but Overrated

An "exome" is the technical term for the part of a DNA sequence that matches the genes in the genome. Or, put differently, all DNA that is not considered to be junk DNA (see Chapter 3). At the moment it is easier in a clinical context to conduct an "exome analysis" rather than a complete genome analysis. One of the reasons being that it costs less. At the same time it is also because the genes are the regions that researchers in this day and age understand the most of and that enable an easy distillation of relevant information.

It is definitely a good idea to use this technique for diagnosing patients that have unknown genetic deficiencies. It would also be useful if these tests were reimbursed by health insurance. But doctors, patients, and parents of children with an undefined genetic condition have to realize that exome analysis identifies the genetic cause of disease in slightly less than 50% of studied cases. This is partly due to technical limitations, but it is also becoming clear that the "classic" mutations, as described in this chapter, do not explain all genetic defects.

by copy number variants is long. But again, we are dealing with a spectrum in which most copy number variants contribute to the normal variation between humans, and that includes variation in talents, characteristics or predisposition for disease.

What Will Reading Genomes Change in Genetic Diagnostics?

The new appliances for genome research have an enormous capacity. Researchers could read all of the genes of a patient simultaneously, or analyze a couple of genes for tens or hundreds of patients at the same time. The latter approach is used when it comes to screening the breast cancer genes and pushes down the cost of the test per patient. The former is useful in a case where the patient has a disorder that is more than likely a genetic disorder according to the clinical geneticist or other medical specialist who made the diagnosis, but when they have not yet pinpointed which gene holds the cause. There are a couple of rules of thumb that geneticists use to determine when it is appropriate to use a genome-wide test. Clinical insight and experience with genetic disease play an important role, as well as considerations about the optimal use of the money from the health care system to apply expensive tests to some patients and possibly not to others.

There is a practical approach that often combines both: a lot of disorders are genetically-speaking heterogenic. Different genes may be linked to the same disease, but there are so many of those genes for a given disease that it is practically and economically less interesting to use classic Sanger technique. A good example is the inherited hypertrophic cardiomyopathies that can be caused by a mutation in one of many genes linked to the contractile, structural and electrical machinery of the heart. With "massively parallel sequencing" the mutation scanning of a panel of genes can now

Epigenetics or How Genes Cannot Be Separated from the Environmental Factors at Play

In Överkalix, an isolated community in the north of Sweden, chroniclers have data that goes back decades and is extremely detailed about harvests, births and deaths. This is ideal data to study the connection between food supply and life expectancy. This showed that boys who grew up in a time with sufficient food supply would have grandsons who had a higher risk of diabetes and heart problems and had a lower life expectancy. The abundance of food had no effect on their granddaughters, which means that the effect is passed on through the male bloodline. This example is an illustration of epigenetics: the genome appears to 'memorize' certain environmental effects that it has been exposed to.

Epigenetic changes in the DNA cause a genetic modification that is passed on to future generations. This change is not included in a change in the order of the As, Cs, Gs and Ts in the DNA itself, so cannot be seen as a classic mutation that changes these bases. However, it is the consequence of a chemical modification or lockdown of the bases in the DNA itself. This causes differences in the extent to which the DNA is protected or exposed to regulatory proteins that are bound to the DNA. This way these proteins will influence the readability of the DNA code.

Epigenetics is a normal natural process. This gives stem cells, just like all other cells, all the genetic information they need to create every possible tissue type. Epigenetic factors determine which cells grow up to be blood cells, which ones will be liver cells, etc. As soon as this has happened, specific epigenetic factors will block the genetic possibilities of these cells. Even cancer has an epigenetic component: in this case it is about badly programed cells, which cause them to proliferate uncontrollably. It is in addition to this that epigenetic processes make sure the be done for a reasonable price. This is currently the approach that is used for most 'genomic' tests. But as the technology evolves and becomes less expensive, genome-wide approaches will replace this 'targeted' approach.

Nevertheless, in most countries, the implementation of genomic analysis in clinical diagnostics is slowed down by the fact that it is expensive and it's not yet regulated how health insurance should reimburse the patient. There is hope that this will be fixed soon.

genome has a 'memory'. Research in mice confirms that the environmental factors of a certain generation can have an effect on future generations. In a particular study they gave pregnant mice a diet that changed the fur color of their offspring. Not only would the first generation have a different fur color, but also the ones after them, who were never on that diet. Another example can be found with cloned mice: sometimes their fur is different than that of the original mouse. Their genome is identical but their *epigenome* is not. Similar 'epigenetic' variants occur between identical twins: their genetic code is identical (see Chapter 2) but not their epigenome.

“The genome appears to 'memorize' certain environmental effects that it has been exposed to.

When people talk genetics they tend to overemphasize predisposition: if it is in your genes, you cannot escape it. Epigenetics forms the bridge between *nature* (predisposition) and *nurture* (education and environmental factors) and shows that not only who you are but also what you do influences your own development and the development of your offspring. To say it with a metaphor, genetics is complex sheet music that holds all the key elements of a classical masterpiece, whereas epigenetics are the musicians and the instruments that bring the music to life.

The study of epigenetics is a relatively new discipline. Preparations for a *Human Epigenome Project* have started, following the analogy of the *Human Genome Project*. It is always risky to try and predict the future, but researchers expect that a better understanding of epigenetics could be useful for instance in the treatment of different types of cancer.

Epigenetics in Practice: The Consequences of the Hunger Winter in 1944

In the harsh winter of 1944, German occupying troops took revenge for a railway strike in the Netherlands in the fall of that year. They blocked the transport of food to the West of the Netherlands. In the first months of 1945 more than 20,000 people died of starvation.

The survivors went through a lot of suffering. Pregnant women had children with a high risk of diabetes, obesity and heart and vascular diseases. The health problems of this first generation after the war can be explained through malnutrition during pregnancy. But also the second generation had the same type of clinical problems, while in the meantime the Netherlands had become a prosperous country with no food shortages. A lasting, hereditary effect is explainable only through epigenetics. It is an illustration of the fact that environmental factors can have an influence that is 'written on (not in) the DNA' and passed on between generations.

5
Angelina Jolie's Choice

In 2013 Angelina Jolie had her breasts removed as a preventive measure because she has an increased risk of breast cancer. Why does she have a greater chance of having breast cancer and how hard is it to make such a decision?

Angelina Jolie is one of the most famous actresses of our time. Through her open letter in *The New York Times* in 2013, she created worldwide awareness of breast cancer and especially the hereditary nature of the condition in some women. It looks like her action had more effect than any other campaign had around breast cancer; suddenly everyone became worried about the matter. Breast cancer is in fact the most common form of cancer in women. In the Western world, almost one in ten women gets breast cancer. In most cases, this happens after the age of 50. But some women face a risk that is much higher earlier in their lives.

Breast Cancer: Hereditary or Familial?

In her open letter Angelina Jolie described how her mother passed away at the age of 56 due to breast cancer. Because of her family history of breast cancer, Angelina had a genetic test done. She had inherited a defect in the BRCA1 gene from her mother.

It is true indeed that between 5% and 10% of women who have breast cancer have a family history that is positive for breast cancer and/or ovarian cancer. Thus, in some families there is a clear genetic predisposition that leads to a higher risk for these types of cancer. BRCA1 and BRCA2 are the two genes that have the strongest association with breast and ovarian cancer. Mutations in BRCA1 or BRCA2 are responsible for the high risk in nearly 30% of the families with a clear history of the disease. This means that there are other families in which breast and/or ovarian cancer are relatively frequent in which the genetic cause has not been identified. In some

Cancer Is Caused by Damage in the DNA

Cancer always originates from genetic defects that appear in the cells of our bodies. We call them *somatic mutations*. They happen in a spontaneous fashion. When cells divide, their entire DNA content is copied to guarantee that every daughter cell has a complete genome. This happens very quickly and is incredibly efficient. But sometimes the results are not perfect. You can compare this to the way monks used to copy books in the Middle Ages: carefully and with an eye for detail, but sometimes a typo still found its way into the copy of the manuscript. Thus, the daughter cell may contain an error that was not present in the mother cell.

Mutations also occur in cells that are not planning on dividing, thus the risk is not limited to cell division. The best example is the damage done to DNA when a cell is exposed to ultraviolet light, radioactive radiation or the harmful substances that appear in exhaust gases and cigarette smoke. But even if you avoid all of the above-mentioned harms, you will still not be completely safe: the chemical reactions in our own cells can also cause damage to the DNA. It all depends where the typo occurs, and whether or not it is corrected eventually. Just like in a cookbook: if the typo does not hit an essential step in the recipe, not too much harm will be done.

Fortunately our cells have "proofreading" systems that do an amazing job. A mistake during cell division is corrected even before the daughter cell is released; cell synthesis is stopped until everything is checked and corrected. Moreover, a mutation might also be detected by the control systems of the cell that seem to constantly scrutinize the content of our DNA. The affected base will be removed from the DNA and replaced by the original. Our cells even possess a system that reattaches the ends of a broken DNA strain when a chromosome happens to break as a result of cell division or irradiation. Warning: if you lie in the sun for a day unprotected there will be thousands to hundreds of thousands of mutations in the skin cells, in other words, you are severely challenging your repair systems. That is why it is strongly advised to wear sunscreen.

Sometimes a mutation escapes the repair mechanisms. When it accidentally affects an important gene, for instance one that plays a part in the growth of a cell or one of the genes that are crucial to DNA repair itself, things can start to go wrong. We say they can start to go wrong because it is clear that it takes more than one mutation before things go "pear-shaped" and the cell will start to divide beyond control. Nevertheless, it is 'molecular accidents' like this that eventually cause the formation of tumors.

The abovementioned BRCA1 and BRCA2 genes, as well as some of the other genes that are associated with an increased risk of familial breast and ovarian cancer, are genes that play an exact role in one of the repair processes: they are involved in the detection of broken chromosomes and their restoration. Thus, we need BRCA1 and BRCA2 because, as long as those two genes are intact and functioning correctly, our cells are well protected.

Some people – and Angelina Jolie is one of them – are born with a defect in one of those two genes. They are missing a working copy of the BRCA1 or BRCA2 gene from the start. They only have one normal copy – remember that we have two copies of every gene, one that we inherit from our mother and one that we inherit from our father. You can say that they start with half of the DNA repair capacity. On the level of the cell the latter statement is not entirely true: there are small compensation mechanisms, also in cells from people with a BRCA1 and BRCA2 mutation but that is a minor detail; the evidence is that they have an increased risk of cancers, which means that the system is more prone to failure in their cells.

of these cases, other genetic defects are found that are responsible for the high risk. But not a single gene in these series affects as many families, and thus causes as many cancers, proportionally speaking, as BRCA1 and BRCA2.

Thus, in several high risk families researchers are not able to find a *prime suspect*. This brings us to the conclusion that in these cases different genetic defects –or even combinations of relatively frequent genetic variants– make a small contribution to the increased risk of cancer in those families. Nonetheless, in 80% of breast cancer cases there is no hereditary or familial explanation.

We expect that in the near future not only will genetic testing for BRCA1 and BRCA2 and the other 'high risk' genes be offered, but also tests for the more moderate risk factors. The problem with the latter is that it will be a lot harder to precisely predict this risk, and thus much more difficult to adequately counsel the individual women, not least because 'bad luck' also plays a role in the incidence of cancer.

Reduced Resilience Can Lead to Tumors

People who are born with a defect in the BRCA1 or BRCA2 genes, are at high risk. Indeed, when a mutation – what we will refer to in the text box as a somatic mutation – affects the healthy copy of the BRCA1 or BRCA2 gene in their breast tissue, all of a sudden they do not have a *back-up* copy. Thus, in this breast cell, BRCA1 and BRCA2 cannot function normally. Specifically, as of that moment the cell is not capable of efficiently repairing damage to chromosomes. It is not long before the cell will encounter problems. Either the cell will die or a tumor will start to grow.

The risks connected to an inherited defect in these genes are great: women with a defect in the BRCA1 or BRCA2 gene have a 60 to 80% chance of developing breast cancer before the age of 80. Their genetic predisposition often causes the cancer to hit them at a younger age. Women without a family history of breast or ovarian cancer have a 3 in 1,000 chance of getting a tumor before the age of 35. In the case of a BRCA1 or BRCA2 defect this increases to 1 in 100. Besides, they also have a 10% to 40% chance of getting ovarian cancer, depending on the nature of the defect. Exceptionally, there are people who inherit a defect in both genes, BRCA1 and BRCA2. You can imagine that their risk is even greater.

> **"If Angelina Jolie had not inherited the defect in the BRCA1 gene, her risk of having breast cancer would be the same as that of any other woman.**

It is not known why defects in the BRCA1 and BRCA2 genes typically cause breast or ovarian cancer. Analogous observations occur for other repair genes. For instance, mutations in the genes that mainly play a role in detecting and correcting chemical modifications in the DNA are more likely to cause bowel or uterine cancer. It may be due to the type of exposure that these cells get to carcinogens.

Overall, cancer is a condition that is not easy to grasp: different genetic defects cause different types of cancer, and the more the matter of cancer is studied, the more researchers are amazed by the complexity of the disease.

Sometimes the Cancer Risk Runs in the Family

Predisposition for hereditary breast cancer – like in Angelina Jolie's case – results from the inheritance of an *autosomal dominant* condition. We have already brought up the reason: You receive a copy of each gene from each parent. If one or both carries the defect in the BRCA1 or BRCA2 gene you have a 50% chance that you inherit the defective copy. Angelina inherited her mother's defect. Her brother also has a 50% chance that he inherited this as well. If he carries the defect he has a greater risk of having prostate cancer and breast cancer. Men can also get breast cancer but the absolute risk is low compared to women's risk.

When conducting a consultation in a clinical genetics center, clinicians will first ask about the patient's family history and document that thoroughly. As in the case of Angelina Jolie, they will draft a family tree and see if any of the family members suffered from breast or ovarian cancer (for men it is prostate and breast cancer). After Angelina Jolie's mother was diagnosed with breast cancer, undoubtedly such a family tree was drawn up with all the relevant information. In practice, the search for the mutation in the BRCA1 or BRCA2 gene starts with a family member that had cancer. This led the clinical scientists to find out that Angelina's mother carried the defect in the BRCA1 gene. Once the familial mutation is found, other family members can now be tested. This is called predictive or presymptomatic testing.

Angelina had a one in two chance of inheriting this defect from her mother. Because she inherited this defect, she has a greater chance of getting cancer. If Angelina Jolie had not inherited the defect in the BRCA1 gene her risk of having breast cancer would be the same as that of any other woman.

To Get Tested or Not to Get Tested?

Angelina Jolie's chance of getting breast cancer was portrayed correctly in most newspaper articles. What was not correct, however, was how they made it seem that Angelina Jolie was the first woman to have a preventive breast removal. There were many women who already had to face this hard choice. A predictive genetic test can bring clarity about whether someone carries the defect in the BRCA1 or BRCA2 gene. It is a test done to "predict" the chance of a person developing cancer. It goes without saying that this test can only be offered when the hereditary defect in the family is known.

The decision whether or not to get yourself tested in such a situation is a difficult one. It brings up fears and uncertainties linked to the fact that you could or could not be carrying the familial defect that could lead to cancer. This brings to life many existential, medical, psychological and practical questions. On top of that, breasts are strongly connected to a woman's self-image and her experience of sexuality.

In the case of a positive test result (meaning that the familial mutation is present), there are more questions that surface. Will I actually develop breast cancer? Should I opt for regular breast exams so the cancer is discovered at an early stage and treatment can start early? Or should I choose to preventively remove my breasts?

Because of these multiple questions it is important that women prepare well for undergoing a predictive test. This comes with obtaining the correct knowledge about genetic risks and the risk of breast cancer, information about possible treatment, and emotional support. A real treatment does not exist, but women can choose to preventively have their breasts removed. This choice is a medical and psychological heavyweight. Such testing should therefore not be offered without genetic counselling, because the impact of the result is enormous, as was illustrated by Angelina's case.

Family Communication

Angelina Jolie's open letter was an unprecedented and very open way to talk about breast cancer. It was her intention to break the taboo. This can lead people to discuss breast cancer within the family more openly. Indeed, a special element of hereditary conditions is that this information can be relevant to other family members.

Since the genetic information about hereditary breast cancer generally belongs to the confidentiality of the patient's medical file, genetic centers cannot just contact family members at risk. The patients are the key messenger responsible for informing their relatives. Since breast cancer is a quite frequent condition, family members might not be aware of its hereditary character until they are actually informed that a genetic risk factor was found in their family that exposed them to a higher risk of developing breast cancer. Research shows that in most cases direct family members (children, brothers, sisters, parents) are informed about their genetic risk, although male family members are often forgotten. However, men can also carry a defect in the BRCA1 or BRCA2 gene. Suppose that Angelina's brother also inherited the defect and were to have daughters, then the situation for them would be the same as Angelina's, before she had herself tested: a 50% chance of inheriting the defect.

When it comes to distant relatives, family communication goes wrong more often. It is usually about branches of the family that are not in regular contact. Seemingly, a lot of people find it difficult to talk about something personal like breast cancer in these situations with distant relatives. Sometimes family feuds or divorce stand in the way of this communication. Or sometimes people suppose that "the others already know" so as not to have to disclose the bad news themselves... People's and legislators' view on this are changing; take

Oncogenes

Oncogenes are the genes that play a normal part in the development of a human, usually in early, embryonic stage. Most of them stop working once their task has been completed. Sometimes a mutation or a fracture in a chromosome will "wake up" such a gene, in a cell that really has no need for the protein that the gene produces. It is also possible that a protein will have a new function because of a mutation. If this happens with a gene or protein that is responsible for cell growth and cell division, there is a risk that the cell will start multiplying and cause a tumor.

No one can live with a deleterious, inherited defect in an oncogene (a few examples set aside). The presence of an activated oncogene in all the cells inevitable leads to death. But a mutation in an oncogene can also occur at a young age, which explains why children can also get cancer. The risk is low and cannot be predicted in any way. From a genetic standpoint it only takes one mutation in one of the two copies of the gene. It does not matter that there is a second copy that still is normal. The defective gene will start a chain reaction that the normal gene cannot stop. Genes in this category (in which mutations cause one or multiple types of cancer) are BRAF, KRAS, MYC, BCR-ABL, etc.

France for instance, where not sharing such critical information is now a legal offence. The most recent UK guidance states that you can separate an individual's genetic test result as being confidential from the familial risk information as being shared, and that a doctor can therefore contact family members.

Talking about It Is Important

Informing family members is definitely not easy. On the one hand, one feels the moral responsibility to do so; on the other hand, you realize that you are burdening your family members with a medical problem that they are not yet aware of. The role of being the messenger is accompanied by many emotions. You have to talk about your own experience with the cancer even if you would rather prefer forget about it. You could also feel guilty about the pain and grief that your information might cause in others. The messengers are also exposed to the emotions of the people getting the news. Not everyone is happy to hear it. Whoever is confronted with an unknown risk that could impact on their life needs time to process this. If you receive good counseling from your doctor, geneticist or specialized health care professional, you will also be able to seek advice for this.

There are families where everyone is informed immediately, which is good for cancer prevention, especially for those at a high risk. Let us hope that this process runs more smoothly thanks to Angelina Jolie's approach. Recently, some countries (such as Australia) have also enacted legislation to allow healthcare professionals to break their medical confidentiality under strict conditions if a patient refuses to communicate risk information to relatives, or that legally obliges patients to share the information (such as in France).

Tumor Suppressor Genes

Tumor suppressor genes are involved in repairing DNA damage or in suppressing the effects of other genes. In this case the cells do not have a lot of problems if one of the two sets of genes is turned off because of a mutation: they still have a back-up (the healthy copy of the gene on the chromosome inherited from the other parent). But when this one is also turned off as a result of a 'somatic mutation', the cell will have a problem (see above). In this chapter we gave the example of the BRCA1 and BRCA2 genes. These are tumor suppressor genes that repair chromosomes and protect cells. Other genes in this category are NF1 (mutations in this gene cause neurofibromatosis), RB1 (responsible for retinoblastoma), TP53 (causes the Li-Fraumeni syndrome), etc. Someone can be born with a defect in those genes and still develop normally, but will have a greater risk of getting cancer. Some of these conditions, known to be familial cancer syndromes, will have tumors appearing at a young age. TP53 is one of the most fearful genes, while BRCA1 and BRCA2 rarely cause tumors before the age of 20 or 25.

Angelina Jolie and Her Children

Angelina Jolie and Brad Pitt have three biological and three adopted children: the twins Knox and Vivienne, Shiloh, Zahara, Pax and Maddox. Currently, all are minors. The biological children Shiloh, Vivienne and Knox have a 50% chance of carrying the defect in the BRCA1 gene just like their mother.

Technically speaking, these children could also be tested for defects in the BRCA1 gene. Healthcare professionals and many geneticists are usually not in favor of conducting these predictive tests on minors. A survey of European geneticists showed that 96% of respondents would rather not test a six year old. The main reason for the refusal is that the test does not hold a medical advantage at that age. A predictive genetic test is useful only if it can lead to medical treatment or preventive measures that could prevent, slow down or postpone the development of a condition during childhood. For other types of cancer, like retinoblastoma (cancer of the retina) and some aggressive types of bowel cancer, specialists do perform these tests in minors because they can affect the child at a young age.

Another reason why specialists hesitate to test children for a condition that appears later in life is an ethical issue about autonomous decision-making. When parents decide for their children to undergo such a test, they remove from the child the chance of making that decision later in life. Particularly when there are no medical advantages of testing at a young age, one has to wonder whether the advantages of testing really outweigh the potential harms.

Waiting a Little Longer Will Often Be a Better Solution

Minors themselves, more than their parents, will have to find a way to live with the consequences of the test results, no matter what. That by itself is a good reason to let them make the decision whether or not to undergo a predictive genetic test. This way the genetic information remains confidential. Besides, when you offer the test at a later age the person is usually in a better psychosocial environment when they receive the results. A predictive genetic test takes place in a multidisciplinary context. Before the test, the

applicant will receive all the information regarding the condition, about the test itself and about the possible psychological impact of it. This multidisciplinary approach is necessary to guarantee the wellbeing of the test requester and to help prepare them for how to handle its possible outcome. Therefore, guidelines suggest postponing these tests until minors have reached a level of maturity and competence to understand the nature of the decision and its implications.

Cancer and Taboo

Mary-Claire King, a geneticist and mathematician from Berkeley, California, started in 1974 to research familial breast cancer. In 1990 she had gathered genetic evidence to show that there had to be a gene which, when mutated, causes cancer on chromosome 17 but that this gene would play a role in only a fraction of the affected families. She called the gene BRCA1. She had no idea what the gene would look like: it was 1994 before the DNA sequence of the gene was decoded. Mary-Claire King was one of the pioneers of the genetic research towards finding the cause of breast cancer. She received international appreciation for this in many ways. In 2006 she received an honorary doctorate at the University of Leuven. During the ceremony she talked about how in the early years, it was nearly impossible to conduct genetic research for "female" cancers, or even gather family data and DNA samples of the patients. That was because the disease was surrounded by an enormous taboo, with doctors as well as with patients. Certainly, this taboo was not limited to the United States, where she conducted her research. It was one of the reasons why the BRCA1 and BRCA2 genes were discovered relatively late. The other reasons were that it is extremely difficult to identify the causal genes when the disease is very frequent and presents in multifaceted ways and that they were genes of previously completely unknown function that could only be traced using the most modern genetic technologies.

In 2013, a movie was made about Marie Claire King's quest to find the first breast cancer gene: *Decoding Annie Parker*. The producers portray Annie Parker, a women suffering from breast cancer with an obvious family history of the disease and with strong willpower to promote scientific research on breast cancer. The movie is a well-deserved tribute to this strong woman. Unfortunately, it has not been widely distributed and, as far we could find out, did not make it to Europe.

6
Give Me Back My DNA!

Whoever undergoes a genetic test should be well informed about the goal of the test beforehand. But what if someone's DNA is being analyzed without his or her permission?

Article 5 of the Unesco Declaration on the Human Genome and Human Rights states that "in all cases, the prior, free and informed consent of the person concerned shall be obtained" and that "research, treatment or diagnosis affecting an individual's genome shall be undertaken only after rigorous and prior assessment of the potential risks and benefits pertaining thereto."

Informed consent can be specified as having an *information* component and a *consent* component. Before a genetic test can be conducted, it is necessary that you are *informed* about the possible outcomes and what that could imply: its goal, risks, medical implications, and the psychological, emotional and financial consequences. This information will help you to consider your options and enable you to consent or not consent in a thought-out way. Depending on the goal of the genetic test, the result can have different clinical, social or psychological consequences. There needs to be enough time and care put into examining whether or not you comprehend all the information given to you. The doctor or medical professional who gives you the information could ask you to use your own words to describe the information you received. This is often not an easy task for people that are not familiar with genetics. Several empirical studies have shown that people may encounter difficulties in understanding and assessing genetic risk and patterns of inheritance and may experience difficulties in processing information about risks.

"The patient has the right to consent freely, before the intervention takes place and after the proper information has been given.

The *consent* component refers to both a voluntary decision and an authorization to proceed. A genetic test cannot be conducted when you are put under pressure, whether it is pressure that comes from family, friends, a partner, a healthcare professional, an insurance company or an employer. Plus one needs to offer the patient the right to change their mind at any time, whether that is not to take the test or not to want to know the results.

It is also important that individuals are not forced to make a decision in a short period of time. For genetic tests with a large psychological impact, for instance a test to see if a person has Huntington disease in a family in which this disease appears, it is especially important that you have enough time to make your decision.

Genetic Testing without Permission

Although there are strict rules under which a DNA test can be performed in a clinic, we have seen in the past various cases of non-consensual testing. A cigarette butt, a hat, a glass, a fork... We are constantly leaving traces of our DNA. Usually this will not have consequences. In exceptional circumstances, for instance with a police investigation, this could hold interesting traces leading to the culprit. But there are clues that "DNA theft" could become a problem

Particularly famous people risk becoming victims of DNA theft. Some fans – or stalkers – just want to collect everything that has to do with their idol, paparazzi are going after that next big scoop and criminals are not scared of blackmail. For example, two newspapers reported in 2002 that the tabloid *News of the World* had tried to collect some hair from Prince Harry (who was 18 at that time) for a DNA test in order to check whether Prince Charles was Harry's biological father. The Crown Prince's secretary was tipped off about the story and took it to Scotland Yard to end this conspiracy. Despite the claims of the newspapers, it needs no further explanation that that the spokesperson of *News of the World* denied the whole conspiracy. The British Court and Scotland Yard could only say, "No comment."

"Particularly famous people risk being victims of DNA theft.

Paternity Tests

But even private events seem to be enough of a reason for some people to conduct a genetic test without the permission of the person involved. It happens that underwear or bed sheets will be sent to a lab to try to prove that someone is committing adultery. A self-respecting laboratory will obviously decline such requests. Laboratories that specialize in forensic analysis can handle these types of investigations, but will not do so without a legal mandate.

Fathers who question their fatherhood might try to have themselves and the child involved tested, or try to get hold of the DNA of the alleged father. Even in this case, if something like this happens in secret, most commentators and legal experts would consider this unethical or illegal.

There are a lot of other situations where DNA is stolen. Within the context of arranged marriages, minors are sometimes tested to see if their future grandchildren will face a greater risk of having hereditary autosomal recessive conditions. Criminals could also use DNA traces of other people to their advantage and try and lead the police and justice system away from them. In the movie *Gattaca* the main character uses falsified DNA to give himself a new identity.

The Hunt for Genetic Material of Famous People

There have been cases in the past where secretly collected DNA of VIPs led to controversy. A couple of examples:

In 2008 someone offered the plate, cutlery and a half-eaten waffle that President Obama used in Scranton, Pennsylvania on eBay using the following sales pitch: "100% authentic, his DNA is on the cutlery". The seller hoped to sell it for over 10,000 USD that he would eventually donate to the Democrat party. But the person who ended up buying it only paid $0.99 plus $6 for shipping.

Your dentist will not agree with the following, but apparently flossing your teeth can be dangerous. Kirk Kerkorian, the well-to-do owner of the movie company Metro-Goldwyn-Mayer, did not want to pay alimony for his daughter that he had with his ex-wife Lisa Border, whom he dated for ten years and was married to for a month. Using a short string of floss that he had genetically tested, his attorney showed that movie producer Steve Bing was the father of the girl with 99.993% certainty. Bing, who was already involved in a trial about the fatherhood of Elizabeth Hurley's son, countersued Kerkorian and his coworkers for DNA theft.

When Britney Spears shaved her head bald at the height of her fame, her hair – and undoubtedly that of many others – was offered for sale on different online auction sites. Even earlier in her career, the trash she threw out was a goldmine for online sellers: paper tissues, a used towel and a piece of soap, cigarette butts and, most of all, pieces of gum. A proud owner offered part of it for sale with a minimum bidding price of 1,000 USD using the sales pitch: "You can conduct a genetic test to determine its authenticity".

The characters have to identify themselves through a finger prick where they take a small blood sample. By falsifying this test the main character is able to pass himself off as someone else. Inspiration is often not that far fetched when it comes to DNA theft.

Ordering a Test Online

The internet has made it easier to buy whatever you want without threshold, genetic tests included. The possibilities and what is useful are explained further in Chapter 21. The question is if the results of genetic tests are stored safely on the internet. A *New Scientist* journalist managed to fairly easily *hack* his colleague's genome without the company even questioning the origin of the DNA sample. His colleague was notified in advance about this attempt. An American government body mentioned this problem in a report. They used actors and were able to show that some internet companies were prepared to conduct a DNA analysis as a "present for the fiancée" of the person that was sending in the sample.

Paternity tests are particularly popular in "secret" test company advertisements. To cross the line even further, some companies mention where to find useful DNA like toothbrushes, Band-Aids, hair, cotton swabs, etc.

Permission and Privacy

Genetic testing is a private matter. The test can reveal medical information that the family member in question does not even know of, or that he or she does not necessarily want to share with others. It can also lead to exposing family bonds or relatives. People who have someone else's DNA tested without their knowledge or consent are invading that person's privacy and autonomy. This includes the right with whom to share the information, but also the right not to know certain information.

In its Human Tissue Act, the United Kingdom has set DNA theft to be a criminal offence. The law has envisioned a three-year prison sentence, a fine, or a combination of both. In most countries, legislation will limit itself to the instruction that a genetic test may not be conducted without informed consent. In Germany and Austria, people have been convicted of unauthorized DNA theft and DNA analysis.

7
We Are All Carriers

We all carry genetic defects that could possibly lead to disorders in our offspring. Should we trace these defects before a possible pregnancy?

We all have a dozen (or more) defects or mutations in our genes, of which most are not harmful because they affect only one of the two copies of a gene. Individuals with those defects ("carriers") usually do not realize that they have these mutations because they do not exhibit symptoms. But if both copies of the same gene contain a defect, individuals might develop symptoms. These are called *recessive conditions*.

Children regularly develop inherited conditions for which their parents carried a mutation, but of which they were not aware. In this case, both parents are carriers and are heterozygous for the genetic defect (to use a technical term). The child with the disorder is homozygous for the mutation in the gene (or compound heterozygous, if the mutation is not exactly the same).

Testing Parents before or during Pregnancy?

In Western Europe one in every 3,000 newborns has cystic fibrosis, the most common life-threatening hereditary disease in the region. Despite the technical possibilities of finding out whether someone is a carrier, it appears that more than 85% of children born with this disorder are to couples who are not aware of the hereditary risk that runs in their family. This is because approximately one in 30 people carry a mutation in the cystic fibrosis (CFTR) gene, thus carrying the disorder. As carriers don't present symptoms, this defect might pass on from generation to generation unnoticed. Couples who carry the defect are usually not aware of their increased risk until they unexpectedly conceive a child with cystic fibrosis.

If these couples are confronted with the news that they both carry a mutation in the CFTR gene, they are faced with a tough decision regarding possible new pregnancies. It forces them to seriously reconsider their wish for children and take reproductive decisions.

“We all have a dozen (or more) ‘defects’ or ‘mutations’ in our genes.

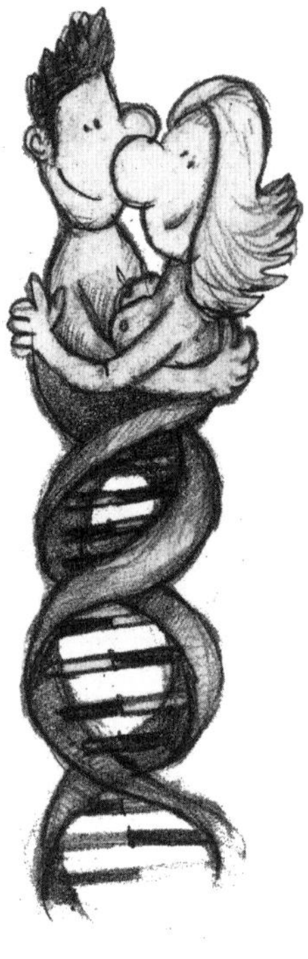

Research shows that the majority of parents with a child who suffers from cystic fibrosis choose to avoid having a second or third child that will also suffer from this disease. These couples can opt for a prenatal diagnosis, followed by pregnancy termination if the fetus has the mutation in both copies of the gene. They can also choose not to conceive any (more) children, choose adoption, gamete donors or to proceed with embryo selection through preimplantation genetic diagnosis (i.e. the genetic diagnosis of an embryo before implantation).

Carrier testing can be done on the parents of a child with a hereditary disorder, but also on siblings, and depending on the inheritance pattern also on other family members who might be at risk of being carriers. If those relatives turn out to be carriers, the next step is that their respective partners are screened for possible mutations in the relevant gene. This approach is called *cascade screening*. Just like a waterfall, the continuity of family bonds is followed to inform people about a hereditary risk that is present in their family.

Carrier Screening as a Standard Procedure

Besides tracing carriers in families in which a certain autosomal recessive disorder is present, it is also possible to actively track carriers in the general population. As described above, most people are not aware that they carry certain mutations and most couples are unaware of the risk they face passing on this hereditary disorder. There are a couple of countries or regions where carrier tests are offered on a regular basis. In Northern Cyprus, for example, they had many cases of thalassemia, a hereditary blood disorder that causes the body to not produce sufficient hemoglobin in a consistent manner. Hemoglobin plays a part in the transportation of oxygen in our bloodstream. One in seven inhabitants of the island carried mutations for this disorder. As of 1970 the government considered the identification of carriers of this mutation to be a national health priority, as an effort to alert couples of their increased hereditary risk and to lower the number of births of children with thalassemia. By 1983 the Greek Orthodox Church actively participated by requiring every couple that wanted to get married to present a certificate stating that they had taken a carrier test.

Sardinia also has a screening program for thalassemia. In some other European countries around the Mediterranean Sea, like Italy and Greece, carrier tests for thalassemia and sickle cell disease are also regularly offered to couples before or during pregnancy. Sickle cell disease, just like thalassemia, is a disorder that affects the way oxygen is taken up in the bloodstream and delivered to body tissue.

In the village of Volendam (the Netherlands) couples planning a pregnancy can get information about four common autosomal recessive disorders. These disorders are more common in that location than in the rest of the Netherlands. The village is made up of a tightly knit community and traditionally people end up marrying someone from within the community.

A Greater Risk in Closed Communities

In relatively closed communities, autosomal recessive disorders will appear more frequently. The effect is not limited to regions or countries where marriage within a certain community occurs more often (see Chapter 10). To give another example: in the past it was uncommon for people living in mountainous regions to migrate. There are specific mutations that find their origin in patients from one specific valley. The Swiss 3905insT mutation in the cystic fibrosis gene is one example. This "founder-effect" will lead to greater risk of carrying the mutation. The same happens in other areas where people used to prefer to marry within the same village. This phenomenon does not restrict itself to cystic fibrosis as it has been documented for other genes as well.

A similar story goes for the Ashkenazi Jews, a cultural and religious group in which metabolic disorders like Tay-Sachs disease and Canavan are very frequent (approximately one in five thousand). In the rest of the population these conditions are extremely rare. This can be explained by the presence of a mutation in the corresponding genes combined with generations of marriages within the community. Jewish couples are now commonly offered carrier tests for ten to fifteen of these severe and untreatable conditions.

Different Rare Disorders with a Large Impact

The examples show that the carrier frequencies differ from population to population, just like the mutations that appear in the different populations. And although each of the autosomal recessive disorders is relatively rare on its own, the combined impact on our healthcare system is rather large. Out of the more than 6,000 known monogenic disorders, approximately 1,100 are inherited

The Pingelap Atoll

On the atoll of Pingelap in Micronesia, an exceptional form of night and color blindness (achromatopsia) appears very often – nearly one in every ten inhabitants suffers from this disorder. The cause is known: in 1775 the population of Pingelap was decimated by a typhoon. Nearly 20 of the island's inhabitants survived this natural disaster, one of which happened to carry a specific mutation. Through the years the carrier frequency on the island has increased to almost 30%. There is a specific term for this phenomenon: the "founder-effect". In this case the origin of the expression of the genetic variation on the island is due to a natural disaster, which stands out because of the presence of this one mutation that appears frequently.

in an autosomal recessive pattern. According to estimates these are responsible for 20% of children's deaths and 10% of children's hospital admittances.

The desirability and feasibility of carrier screening for diseases like cystic fibrosis, sickle cell disease and thalassemia have been thoroughly examined in many different countries. These studies show that positive attitudes towards a carrier screening offer before pregnancy exist among different stakeholders, including couples planning a pregnancy, patient groups and healthcare professionals.

Extensive Screening Is Easier Said Than Done

However, carrier screening programs are not implemented on a large scale. First of all there is a practical problem: many women only visit a doctor or midwife after they have become pregnant, while carrier testing offers the most advantages before the pregnancy. An important barrier is also the population's limited knowledge of carrier testing. When couples are not aware of these tests, they will not bring them up with their doctor. Moreover, our healthcare systems are currently not prepared to offer a carrier test to every couple.

More fundamental objections against carrier screening have also been raised. According to some commentators, an increased offer would also lead to an increase in pregnancy terminations. Even though the focus of carrier testing lies before pregnancy, various couples might probably opt for a natural pregnancy with a prenatal diagnosis and a possible termination of the pregnancy. However, this could also be considered as the normal consequence of providing couples with more information about their genetic risk and allowing them to make informed reproductive choices (see also Chapter 8).

> **“Although each of the autosomal recessive disorders is relatively rare on its own, the combined impact on our healthcare system is rather large.**

Other concerns about carrier testing are the psychological impact on a person when he or she hears the news that s/he is a carrier, the possible problems with communicating the test results and the possible stigmatization of and discrimination against carriers. These can be partially solved by providing people with good information and giving them proper guidance, supported by legislation that deals with discrimination.

Genetic Tests Are Increasingly Put in the Spotlight

These days genetic tests are increasingly available through the internet. Various companies are also advertising and selling carrier tests online (see Chapter 21). An example of such an offer is the so-called "universal genetic test", which tracks more than 100 different autosomal recessive disorders. Compared with the carrier screening mentioned earlier in this chapter, which is limited to a couple of disorders and focus more on common mutations in specific population groups, the company that offers this test chooses a completely different approach. It tries to map out as many disorders and mutations as possible, making the test as "universal" as possible.

Such a "universal" test also has some downsides. For many disorders that are tested using this method, the question can be asked if it forms an important or severe health problem. Should it be the carrier test's goal to screen for less severe or treatable disorders? Should such a test be limited to disorders that occur during childhood? Certain mutations that will also be detected by such a test will possibly have an unclear meaning. Plus, this test feeds the illusion that all possible risks are covered by it, which is evidently not the case. There is always a small risk that remains.

> **"Should it be the carrier test's goal to screen for less severe or treatable disorders?**

Offering carrier tests through the internet also brings up a number of important ethical questions. What about the increasing number of choices that couples are confronted with before and during the pregnancy? Will these couples be able to process this abundance of questions? Will they be able to make a decision in all liberty? Will positive tests lead to the end of relationships? It also questions where we draw the line and which conditions and mutations are severe enough to allow a termination of pregnancy.

More Extensive Tests Will Be Possible in the Future

Technological progress promises even broader tests. Not so long ago a research group published a *proof of concept* to examine more than 400 genes at the same time. This enables researchers to track more than 400 severe autosomal recessive disorders that appear during childhood. These technological advancements and the commercial pressure to undergo such tests puts the topic back on the regular healthcare agenda with the following question: should we offer every couple planning a pregnancy a carrier test?

8
New Test Tracks for Fetus DNA

Today, a fetus's DNA can be isolated from the pregnant mother's blood in order to screen for Trisomy 21. What are the ethical implications of such a screening?

Researchers have developed a method of deriving fetus DNA from the blood plasma of a pregnant woman. Such test can in theory be done after 6 weeks of pregnancy. In technical terms it is called cell-free fetal DNA, or cffDNA, and this DNA usually come from the placenta. We are talking about short DNA fragments of which 80% are shorter than 200 base pairs. These DNA fragments circulate in the mother's bloodstream. During the first trimester of the pregnancy a maximum of 10% of the cell-free DNA comes from the fetus; during the second trimester this can run up to 20%. Post partum this cffDNA rapidly disappears from the mother's bloodstream.

A New Testing Method beside the Amniotic Fluid Test and Chorionic Villus Sampling

The main advantage of using the fetal DNA from the mother's blood is that the amniotic fluid test and chorionic villus sampling are no longer needed to effectively investigate any possible molecular or chromosomal deficiencies in the fetus. This makes the risk of a miscarriage –which is possible in the other invasive testing techniques– disappear. The risk of miscarriage was estimated to be around 0.5 to 1%.

More specifically this means that a lot of women will no longer have to undergo the invasive amniotic fluid test or chorionic villus sampling in the future. As the new sampling strategy only requires the taking of some blood from the mother, this method is considered a non-invasive prenatal test (NIPT).

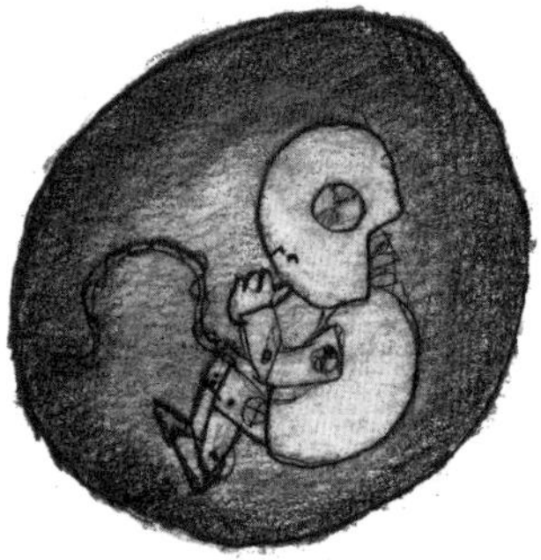

PRENATAL TEST DUMMY

If certain hereditary conditions run in the family, couples can learn about possible hereditary risks and reproductive choices before the pregnancy. In these cases, they will be presented with a classical invasive test. The NIPT has limitations for the analysis of specific mutations in the fetus. Paternally inherited and de novo mutations can in principle be detected, but the technology cannot be used for maternal mutations. At the moment the test is primarily focused on revealing any numerical chromosomal deviations, but the technology is evolving quickly.

In exceptional cases the non-invasive prenatal test can be used to determine the sex of the fetus for medical purposes. For example, in families with severe X-linked conditions, like Duchenne muscular dystrophy, NIPT enables lab scientists to determine the sex of the fetus at an early stage of the pregnancy. If it is a girl, no further testing is required. If it is a boy, an additional invasive prenatal test is needed to determine whether or not the Duchenne gene alteration is present.

> **“It is all about giving couples a chance to make their own, well-informed choice.**

For the majority of women the NIPT will be used to screen Down syndrome (trisomy 21), as well as trisomy 13 and trisomy 18.

At the moment a lot of women undergo a combination test in the first trimester of their pregnancy, in which doctors try and calculate the odds of whether or not the baby will have Down syndrome by looking at the age and the hormone values and by performing a nuchal scan. After the test, the soon-to-be mothers hear if they have a high or low risk of having a baby with Down syndrome. The NIPT

The Discovery of cffDNA

In 1997, Dennis Lo, a researcher at the University of Hong Kong, discovered that fetal DNA circulates in the mother's bloodstream. He demonstrated this by showing that specific Y chromosome pieces of DNA were present in the mother of a baby boy: since a mother cannot have Y chromosomes, the DNA can only come from a male fetus. Later confirmation came from many different sources that the fetus's cell-free DNA is circulating in the mother's bloodstream. After the baby is born that DNA disappears quickly.

The DNA was named "cell free" not only because it "swims" around freely in the mother's blood, but also because in earlier years they found that there were also fetal cells present in the mother's bloodstream. For over 20 years people invested a lot in new methods to trace those fetal cells and perform tests on them. For different reasons it did not turn out to be successful. One of those reasons is that the fetal cells (in contrast to cell-free DNA) can stay in the mother's blood for a while after the pregnancy, which can result in faulty testing during the next pregnancy.

Dennis Lo and his coworkers patented the discovery of cffDNA and the tracing method (see Chapter 22). The American company Sequenom bought the patent rights, continued to develop the non-invasive prenatal test and started commercializing it. The method was siphoned on to the next generation sequencing equipment (see Chapter 2). The test is based on counting the DNA fragments that come from chromosome 21 (and 13 and 18) and a proportionate comparison of the number of DNA fragments with that of other chromosomes. If scientists find relatively more material from chromosome 21, this could be an indication of trisomy 21. The test can be sufficiently trusted and is used around the globe.

The Center for Human Genetics at the University Hospital in Leuven was one of the first academic centers in Europe to have NIPT ready to use. In the United States, many companies that offer the test are fighting each other over patent rights and currently there are a lot of trials pending. A number of those companies also offer the test in Europe. It is becoming a big market.

procedure is currently mostly offered to women with a high risk. It is expected that the NIPT technology will replace the combined testing methods in the future. There are still ongoing studies that analyze how well this test performs in a general population of pregnant women.

A Cobweb of Ethical Discussions

Prenatal screening has always been the topic of many ethical discussions. Some people believe that human life should be protected from the moment of conception. From that perspective an induced pregnancy termination is a non-justifiable act. Others are more open to these procedures and believe that the amount of protection should grow with the embryo as the pregnancy advances.

The goal of prenatal screening should not be to prevent certain conditions: it merely aims to offer reproductive choice to a couple. Some couples will denounce it whereas others will embrace it. The success of prenatal screening should not lie in the number of people who undergo such testing. It is all about offering that choice. If a lot of couples participate, it could mean that few people have space to make their own decisions, whether because of social pressure or a lack of information.

The Pros and Cons of Non-Invasive Prenatal Testing

Non-invasive prenatal testing offers a few advantages compared to traditional testing methods. First of all, the number of invasive procedures and the conjoined risk of a miscarriage decrease. Secondly, the NIPT has fewer false-negative results: fewer women get the news that their fetus has no deficiency when this is not correct. Additionally, if the NIPT could replace the combination test it would be possible to start screening earlier, which would give couples a larger timeframe in which they could choose to continue or terminate the pregnancy. If push comes to shove, the NIPT also allows testing later in the pregnancy.

A fear that arises when it comes to the implementation of NIPT is that it would become too routine, especially when it comes to replacing traditional testing methods. Some say that the social pressure to take the test would increase. After all it is a very easy test. All you have to do is just let a doctor take some of your blood.

A Couple of Medical Terms

An *amniotic fluid test (AFT)* or *amniocentesis* is a prenatal diagnosis in which a small sample of fluid is taken from the amniotic sac through the woman's abdominal wall. After that, the fetal cells from the fluid are analyzed and scanned for any chromosomal defects. Usually an AFT is performed after 14 weeks of pregnancy, but there is a small risk of miscarriage (around 0.5%).

Chorionic villus sampling (CVS) is also a method of prenatal diagnosis in which a tissue sample is taken from the placenta for analysis. This technique enables doctors to track certain chromosomal abnormalities and specific mutations. This test can be done before the first 12 weeks of pregnancy and has a similar risk of miscarriage (up to 1%).

Down syndrome is a mild to severe genetic disorder that affects the intellectual capabilities of a person. Down syndrome also comes with distinct physical characteristics, e.g. a round face. Down syndrome is also called trisomy 21 because it is caused by the presence of a third chromosome on chromosome 21.

Trisomy 13 and *trisomy 18* are genetic disorders that are caused by the presence of a third chromosome on either chromosome respectively. Both disorders come with severe birth defects.

This leaves many questions. Will people be thoroughly informed about the test? Will they be able to cope with the choice if the test turns out positive? The goal of a screening program must never be prevention, but rather to provide people with a choice. But how does one guarantee that prevention will not become a direct or indirect goal by itself?

Fund Test Programs or Use the Money on People with Down Syndrome?

There is a sharp social contradiction in the discussion of prenatal Down syndrome screening. Implementing NIPT only sharpens this contradiction. On the one hand it is emphasized that couples should be given well-balanced information about prenatal testing methods. The information has to enable them to make a free choice. Whilst making this choice, they have to be able to get appropriate counseling and sufficient support.

It cannot be emphasized enough that couples who choose to give birth to a child with Down syndrome must not be stigmatized for their choice and their children should not be discriminated against or excluded in any way. Offering people a choice in the matter cannot be a statement that people with this condition are lesser people.

“There is a clear difference between historical eugenics and current prenatal screening.

On the other hand, skeptics point out that the money invested in the NIPT could better be used to support people with Down syndrome. According to some, prenatal screening is never neutral. The mere fact of offering such test indirectly states that Down syndrome is in fact a serious deficiency and deals with a life that would be better off not being born. Merely suggesting that couples are truly free in their choice is a utopian idea according to some: there would be medical and social pressure not to give birth to a child with Down syndrome.

However, people living with Down syndrome need to be looked at as equal human beings who deserve our solidarity and consideration. What it comes down to is providing them with better support when it comes to education, better domestic guidance and better medical care. There also need to be sufficient means to accommodate them.

Eugenics

In this discussion the word "eugenics" keeps coming back. Some commentators consider any kind of prenatal screening a eugenic measure, and any involvement of the government in financially supporting prenatal screening as inappropriate.

However, using the word *eugenics* in this context is risky. It refers to a dark period in our history in which people with intellectual disabilities had to undergo mandatory sterilization procedures, race laws prohibited interracial marriages, people with disabilities were killed, some races were considered inferior, and marriages between highly-educated people were encouraged.

But there is a big difference between historical eugenics and current prenatal screening. People can make independent choices. The government can in no way intervene or limit your options. The sole goal of prenatal screening is to enable people to have a choice in the matter. Of course it is clear that the freedom of choice must be protected and guaranteed at all cost. A screening offer cannot be reduced to a hidden form of eugenics in which the pretence of freedom would be an illusion.

A New Batch of Ethical Questions

NIPT comes with even more challenges. The current discussion focuses on introducing NIPT for trisomy 21, along with trisomy 13 and trisomy 18. Technologically speaking, it is possible to track other genetic disorders. First of all, X-linked conditions come to mind, such as Turner's syndrome or Klinefelter's syndrome. Additionally, in a research context it has been shown that it is technologically possible to map out the entire genome of a fetus. This development resembles Pandora's box, where new problems are released.

Eugenics

Francis Galton originally developed the concept of eugenics. This British researcher from the 19th century was one of the first to use statistical methods to research human differences and genetics. He was also a meteorologist and designed weather maps. He was related to Charles Darwin and was fascinated by his book *The Origin of Species*. This book was the push he needed to start mapping human differences. He researched the patterns in fingerprints, facial expressions and specific qualities such as intelligence. He used the term *eugenics* in his book *Inquiries into Human Faculty and Its Development* in 1883 as a description of his research activities. He was always pleading to investigate all elements that could provide insight into making humans stronger or weaker. He went further than only research: in his book he is a firm advocate of early marriages between individuals of a higher social class to generate the best offspring.

Following Galton's legacy, different countries introduced eugenic measures in the early 20th century, including mandatory sterilization procedures, marital limitations, race discrimination, mandatory abortions, forced marriages, etc. Many of the policies in place at that time would be seen as unethical today. The absolute horror of Germany's Nazi ideology in which they used race theories to justify the mass murder and other practices led to the contempt of the eugenic movement. An example of the negative outlook of eugenics today can be found in the *Charter of Fundamental Rights of the European Union,* in which eugenic practices are explicitly forbidden.

When are conditions severe enough to be taken into a screening program? What about mild conditions? Can we come to an understanding when it comes to those? Attempts to make a list of severe genetic disorders for inclusion in prenatal testing have been unsuccessful in the past because the severity of the condition was estimated differently. Should we include conditions that could benefit from early treatment? Would this be for treatments that start during pregnancy or just after birth? Would it also be the case if these conditions see the light of day later in life? How useful is it to have more information about the fetus? Is knowing more necessarily better?

Choosing Sex

NIPT can also provide us with information about the gender of the fetus at an earlier stage. Some people fear that this is going to lead to a higher number of abortions because of the fetus's sex. In most countries it is forbidden to do so for non-medical reasons, but it is very hard to check that. Most Western countries have legislation that allows abortion during a specific, legally determined period. In Belgium, for example, a woman can have an abortion up to 12 weeks after conception. In the Netherlands it is 24 weeks.

There can be different reasons why a woman chooses to end her pregnancy: health issues, age, family circumstances, socio-economic conditions, etc. There is an obligatory waiting time, but abortion centers do not evaluate the reasons why women have abortions. If a couple was actually having an abortion because of the fetus's gender, they might easily hide their real motive.

Sex Determination as a Demographical Time Bomb

Moreover, in some strong cultural or social settings couples might prefer to have a boy. A study from India shows that the number of girls has been consistently dropping in recent decades. In 1981 the number of girls between 0 and 6 years of age was 962 per 1000 boys, in 1991 this was 945 and in 2011 just 914. This is a tragic scenario. In India alone, over 10 million girls were never born. Numbers vary, but the same problem is seen in other Asian countries and amongst migrant groups in Western countries. Over time the abundance of males could cause dramatic social and societal problems

Unfortunately, implementing NIPT could further strengthen this phenomenon. This is the reason why some laboratories do not generate the sex in their analysis or do not report the sex of the fetus. But would this solve the problem? Anyone who really wants to know the sex of the fetus only needs to undergo an ultrasound or take a blood sample and to send it to one of the many internet companies. Couples who really want a child of a specific sex will stop at nothing. The only solution is to abolish the cultural and social factors that cause gender selection.

Paternity Tests before Birth

Companies that offer NIPT over the internet sometimes combine NIPT with paternity tests. Besides a blood sample of the mother, you need a biological sample of the suspected father or fathers. Some companies are willing to perform DNA analysis of bodily fluid such as saliva or sperm, or even a toothbrush, piece of gum, hair, etc. Permission of those involved is often not required by those laboratories. This process could lead to abortions because the baby was conceived with an unwanted male partner. Again, we come across ethical and legal dilemmas that don't hold a quick, all-inclusive answer.

9
Designer Babies

In-vitro fertilization makes it possible to genetically screen an embryo before implantation in the uterus. What is a responsible use of this technology?

If there is a hereditary, monogenic disease that runs in the family, couples face a difficult decision. Either they are dealing with a dominant disorder, in which one of them has the condition and in which case genetic testing can show that he or she will develop the disorder. In that case the parent might pass this genetic defect onto the next generation: for each pregnancy there is a 50% chance. Or they are dealing with a recessive disorder and both parents are carriers. In this case the risk of passing on the disorder to the next generation is 25%. In the case of X-linked, recessive disorders, every boy that is born will have a 50% chance of having the disorder (see Chapter 23 for details on X-linked inheritance).

Enough Decisions to Be Made, but They Are All Difficult

In these situations, some couples decide not to have children, whereas others opt for a pregnancy combined with prenatal diagnosis. Through DNA analysis from cells taken from the amniotic fluid (taken through an amniotic fluid test) or from samples collected from the chorionic villus in the placenta (through chorionic villus sampling), it is possible to trace genetic defects in the fetus. In the case of a positive test result, couples could terminate the pregnancy.

> **“Healthcare professionals should be non-directive.**

In other cases couples will choose a pre-implantation genetic diagnosis (PGD). In that case the fertilization will take place in vitro. After the woman undergoes hormone stimulation, clinicians 'harvest' the egg cells and fertilize them with sperm cells on a petri dish. On the third day, when the embryos have 6 to 10 cells, an embryologist will isolate one or two cells per embryo. Through DNA analysis of these individual cells it becomes clear which embryos do not carry the family's genetic defect. These can be implanted in the woman's uterus. A disadvantage of PGD is that the success rate of in-vitro fertilization (IVF) is relatively low (generally less than 20% per cycle).

Some consider prenatal diagnosis and PGD unacceptable because of the sanctity attributed to life from conception onwards. Indeed, with prenatal diagnosis, in the case of an embryo with a disorder, the pregnancy will likely be terminated. With PGD, different embryos are cultivated in vitro before they are tested. This means that deviant embryos will get lost, but that various leftover embryos will remain after the procedure.

Others accept prenatal diagnosis or PGD to avoid the birth of a child with a severe genetic disorder. Usually the older the embryo gets, the more it deserves to be protected. This is why some people have less difficulty accepting the loss of an embryo in a petri dish (like with PGD) than accepting pregnancy termination (after prenatal diagnosis).

Whoever is confronted with prenatal testing and/or pregnancy termination is facing a heart-breaking choice and will never forget such an intervention, no matter what their outlook on life may be.

Good and Responsible Guidance is Important

The task of healthcare professionals when guiding couples facing these difficult choices about prenatal diagnosis and PGD is not always simple. On the one hand they have to provide the couple with sufficient information and give them the liberty to make their own well-informed decisions. Health care professionals should be non-directive: they should not push the decision in a certain direction.

On the other hand, healthcare professionals have a responsibility towards the future child. Especially in the context of PGD, they play an important role in starting a pregnancy. Difficult situations can arise in which the view of healthcare professionals and the view of the parents conflict.

Severe Disorder or Not?

For example, there is sometimes a discussion about the indications for PGD. At the moment PGD is limited to parents who run a higher risk of having children with severe disorders like cystic fibrosis, spinal muscular atrophy, or Duchenne muscular dystrophy. But which other disorders can be classified as 'severe'? It is usually extremely hard to make a distinction between severe and non-severe disorders. Besides, a disease can develop itself in different ways, in all possible variations. Sometimes different mutations form the basis of the disorder. Sometimes it just takes different forms of presentation. Moreover, things become especially complicated when one is dealing with conditions that only develop in adulthood, like Huntington disease or breast cancer. How families experience a disorder varies. Even healthcare professionals have different experiences and views on the severity of hereditary disorders.

These considerations prevented healthcare workers from drafting a list saying in which cases it is acceptable to have PGD. Even for prenatal diagnosis, most countries will not have such a list. It is left to the parents, in cooperation with the healthcare professionals, to determine if they are dealing with a severe disorder.

> **“It is usually extremely hard to make a distinction between severe and non-severe disorders.**

The question, for example, whether it is desirable to perform PGD for breast cancer because of the mutations in the BRCA1 and BRCA2 genes led to a hefty political discussion in the Netherlands in 2008. As we discussed in Chapter 5, 60 to 80% of women who carry a defect in these genes risk developing breast cancer before the age of 80. Debates focused on the acceptability of PGD in the light that 'only' 60 to 80% of the women actually develop breast cancer, that breast cancer doesn't develop during childhood and that interventions like breast removal and removing the ovaries can be done, thus bringing up the question whether breast cancer is severe enough to initiate PGD.

Savior Babies

PGD has already been used to select embryos with a specific tissue or HLA type to bring a child to life that can be used as a donor for stem cell therapy for a brother or sister with a life threatening disease. It is mainly about children with cancer or a form of leukemia with which different approaches were unsuccessful. Often this is due to the fact that they cannot find a matching stem cell donor in the worldwide collection of umbilical cord blood samples. HLA stands for human leukocyte antigen, which is an important part of the immune system and can be found on white blood cells. When the HLA type does not match, the immune system will react and the bone marrow or stem cell transplant will fail. There is resistance to the so-called 'savior baby' because one child is 'created' as a means to save another child's life. According to some people this is using an individual in a commoditized way and cannot be justified. This situation can be looked at differently. That an embryo with a specific HLA type is chosen does not mean that this child will somehow be unwanted or less loved than the other child. The desire for another child can be just as strong as the desire to find therapy for a sick, already-existing child. Many different fertility centers will support the demand for prenatal HLA procedures to find a suitable donor for a sick brother or sister when all other forms of therapy have been considered, there are no other donors to be found in the international database and if the decision is properly motivated.

From Avoiding Hereditary Disorders to "Designer Babies"

The variety in expression of certain disorders, the presence of preventive and therapeutic possibilities and the fact that some disorders only become noticeable when the patient has reached a certain age are all factors that reduce the support for PGD and prenatal diagnosis. Moreover, instead of using the technology to prevent children with severe genetic disorders from being born, some fear that parents will use it to select certain genetic traits that they want their children to possess. In this context the term 'designer baby' is often used. Would-be parents would select a list of traits that they want their children to have. For example: the color of the eyes, hair color, body length, sex, but also traits like intelligence or temper.

In a lot of discussions about designer babies there seems to be a lack of scientific reality. A lot of the abovementioned qualities are multifactorial: it takes a lot more than just one genetic component and they are usually influenced by the children's environment. Usually it is technically impossible to choose a specific trait.

Selecting for a Hereditary Disorder

The discussion about the use of these methods for the purpose of selection is still ongoing. For example a conflict arose when a number of deaf couples wanted to use PGD explicitly for the purpose of conceiving a deaf child. There were similar cases involving dwarfism. In these cases the use of PGD was not to prevent children from being born with certain severe hereditary conditions but on the contrary, to select a child with a specific trait. The deaf couples in these situations do not see deafness as being a severe condition. They rather view themselves as members of a specific community that uses a unique form of communication. Some see deafness as a form of human variation, just like having a darker skin color or being homosexual. They plead to allow PGD in this context as a form of reproductive freedom.

No Support from Public Opinion

In practice it seems like public opinion does not support this choice. Most people feel that it is against the minor's best interest to choose an embryo that has a significant risk of developing a severe genetic condition or limitation above an embryo that does not have this risk. It also comes from the idea that deafness is not just a variation but holds clear limitations. For example, it is found to be more difficult for deaf people to participate in social settings. Reproductive freedom in this context has a heavy impact on the lives of the future children. The possibilities of the future children are limited beforehand.

Deciding the Child's Sex, a Tough Talking Point

With the help of PGD scientists can also decide the sex of embryos. In the case of a severe hereditary disorder that affects only males (for example, Duchenne muscular dystrophy) it is possible to implant only female embryos. The use of PGD for non-medical purposes is a talking point of the ongoing discussion. The argument of *family balancing*, creating a 'gender balance' in children's sexes, is sometimes used as an argument to allow deciding the child's sex through the use of PGD for social reasons. A couple of fertility centers in the United States offer this procedure. This approach has to endure a lot of criticism because it touches human dignity and degrades children to being mere tools of their parents' wish list. PGD for non-medical purposes is not accepted in Europe. It is not possible to start fertility treatments just because a couple prefers to have a son or a daughter.

Exclude Carriers?

Are couples allowed to use PGD to make sure that their future grandchildren do not inherit hereditary risks? A couple of fertility centers had to deal with this question in the past. More specifically it is about parents of whom one carries an autosomal recessive disorder or of whom the male is a carrier of an X-linked, recessive disorder. In the first case there is one chance in two that the offspring will be a carrier of the disorder. The chance of this carrier meeting another carrier of the same recessive disorder is relatively small. In the second case, all daughters will carry the X-linked disorder. These daughters have a 50 % chance of having sons who will develop the disorder. Mainly in this second case asking for PGD to eliminate having a child that is a carrier is possible.

Some feel that this should be the parents' decision. Even though the question seems disproportionate because the procedure forces would-be parents to undergo a medically taxing intervention (ovarian stimulation, IVF, implementation in the womb); yet they find it to be the parents' right to prevent their children from carrying the disorder.

MARCEL BREUER BABY

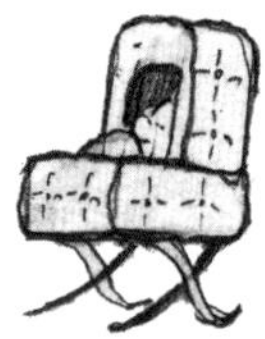

MIES VAN DER ROHE BABY

ARNE JACOBSEN BABY

Others object to this approach. In this case PGD would be used to avoid implanting embryos that are perfectly healthy even though they carry a possible disorder. Being a carrier of a disorder does not make people sick. A medically guided procreation puts a heavy burden on individual couples and on the community. Critics find the burden to be disproportionate to the possible result. According to some there is no responsibility to solve possible reproductive problems of future children.

If a couple undergoes PGD to avoid passing a hereditary disorder on to their own children, there is wider support to select non-carriers over carriers. This comes with the condition that there are enough embryos to choose from. Starting a new procedure of ovarian stimulation to 'harvest' embryos that do not carry the disorder, most health care providers find it too taxing for the couple and too much of a financial burden for society. Others think this decision should be left to the would-be parents.

PGD regulations differ strongly among different countries. In many countries PGD is illegal. The abovementioned dilemmas give a clear image of how sensitive and complex this matter really is.

10
As Long as it Stays in the Family

Marrying within the family is no longer a custom in the Western world, but it is definitely not unusual in the Middle East or North Africa. What are the social and genetic implications of this?

In professional literature a consanguineous marriage is a term for a marriage between relatives: amongst cousins, nieces and nephews and all the possible combinations in between. These marriages are relatively common: some estimate more than 10% of the world's population is either married to a cousin, niece, nephew or is a child of related parents. Demographically speaking, this type of marriage is disproportionately divided over the world population.

A Recent Taboo in the Western World

Among the native population in Europe, North America and Australia, consanguineous marriages are rare nowadays. Historically speaking, this has not always been the case. In a non-globalized society and in geographically or sociologically isolated communities, there once were a lot of marriages within those communities. Besides there would be arranged marriages in wealthy, noble families in order to form political alliances and to protect family assets. There are some well-known consanguineous marriages: Charles Darwin was married to his cousin, just like Albert Einstein, Johann Sebastian Bach, Franklin Roosevelt and Jerry Lee Lewis.

> **"In North Africa, the Middle East and certain parts of Asia the number of marriages between blood relatives can be as high as half of all marriages.**

After the public became aware of the possible medical risks for children, around the middle of the 19th century, marriage between blood relatives became more and more of an exception, even taboo. Governments began to implement legal thresholds. In the Netherlands for example there was a prohibition of marriage between an uncle and his niece, aunt and her nephew, and brother-in-law and sister-in-law (although in-laws are not necessarily genetic relatives). In Belgium, couples in these situations cannot marry, although they can file legal proceedings to lift this prohibition. A marriage between cousins is possible in the Netherlands and Belgium, but is, for example, illegal in some states of the US.

Relatively Normal in Other Parts of the World

In North Africa, the Middle East and certain parts of Asia the number of marriages between blood relatives can be as high as half of all marriages. Social factors often play a role in these cases. Consanguineous marriages often conserve the couple's social relations along with creating stronger family ties and solidarity. Also, premarital financial negotiations are easier to arrange, as assets would remain within the family.

The way religions treat blood marriages varies: within the Roman Catholic Church you need a church dispensation for a marriage between cousins, whereas such a marriage is possible within Protestantism, Judaism and Islam.

Due to migration, consanguineous marriages are making more of an appearance in Europe's regions. There are no exact numbers, however. Despite the lack of figures, a Dutch study showed that approximately 25% of Turks and Moroccans are married to a family member. They also estimate that about 12% of marriages are between cousins. In the Netherlands a discussion recently started about a possible prohibition of marriages between cousins. This discussion was sparked off by the high number of consanguineous marriages in the Turkish and Moroccan communities.

Related Genomes

The closer the family members are in the family tree the more genetic material they have in common. Remember that a child has doubles of each gene – one from their mother and the other from their father – thus creating a genome. Because genetic information is passed down in this way, siblings have approximately half of the same genetic information – they do not have exactly the same genes because they inherit different genetic combinations. Extending this further, nephews and nieces share about 25% with their aunts and uncles, and first cousins share approximately 12.5% of their genetic information.

Having genetic material in common means identical by descent in this context, not to be confused with the percentage of genetic material that all people have in common – which is around 99.9%. Marriages are considered to be consanguineous if more than 1.56% of the genome is shared by descent. To give you an example, this is what your grandson would have in common with the granddaughter of your aunt. Here is how you would calculate it: your grandson shares one-fourth of your genes, you share half of your genes with your mother so a quarter with your aunt, who shares 25% of her grandchild's genome. This being said, it does appear to be relatively distant family relations: those two people probably do not even know each other, let alone know that they are related when they first meet. You would have to piece together your family tree to find out. If you have a large family and everyone lives in the same village, it is not unlikely that people would get married within the same family. But because this percentage is not significantly higher than the part of your genome you share with other non-related people, it is not a problem – throughout history most people kept on living their lives in the same village or region.

Where's the Danger?

We are all carriers of a certain number of recessive defects in genes that, if present in both copies, can lead to severe, recessive disorders. These defects remain unnoticed until two partners who happen to be carriers coincidentally pass them on to their offspring simultaneously (see Chapter 7).

Two family members have a slightly higher chance of carrying the same genetic defect, and if they have a child together there is an increased risk of passing down this recessive, mutated gene. If the family is clearly consanguineous: for example, when these marriages have been a family tradition for generations, then the number of recessive defects per individual rises and the chance of giving birth to a child with a recessive disorder increases.

In general, scientists calculated that a marriage between cousins increases the risk of genetic disorders by 2 to 3% compared to children born from 'normal' marriages. Couples with no blood relations have a 3% chance of having a child who suffers from a genetic disorder – that is a recessive disorder – as opposed to children from blood relatives who have around a 6% chance.

Though we need to point out that this chance is not exceptionally high when you compare it with the percentage of genetic disorders in children from mothers in their forties or older. In these cases we mostly see chromosomal disorders that occur due to issues in cell division. These genetic defects are not the result of consanguineous marriages, but are considered a general phenomenon.

COLUMBUS DISCOVERS INBREEDING ISLAND

In contrary to what someone might expect, recessive disorders do also appear when partners come from different parts of the world, for example when an African man marries a Chinese woman. In both population groups there are mutations that appear in low frequencies in the corresponding genes and an unlucky combination of two different mutations in the same gene will also be expressed with a child bearing a rare, recessive disorder. This chance is smaller to that of members of the same, largely isolated community in which the genetic defect has been passed down throughout the centuries, or in the case of consanguineous marriages. And, as usual, there are exceptions if, for some reason, a certain mutation occurred due to natural selection in both population groups. Think about sickle-cell anemia, a disorder that is ever present in central Africa, but also exists in other regions where there is a high risk of malaria.

Medical Guidance and *Counseling*

A genetic disorder is more likely to occur if it is present in one or both of the parents' medical family history. Blood relatives in a romantic relationship undoubtedly benefit from a genetic consultation to make a reliable estimation of their risk of having a child with a genetic disorder.

There is a high frequency of some extremely rare, recessive disorders with certain consanguineous population groups. The effect is well documented in the case of the Amish, a historically and religiously isolated population group in Pennsylvania, amongst other states. One specific recessive muscular disorder, named *Amish-type nemaline myopathy*, appears in 1 in 500. The same goes for a non-viable form of microcephaly, which is when a child is born with a small head due to a growth defect in the brain, which is called *Amish-type microcephaly*. And then there is also a severe form of epilepsy, named *Amish brittle hair brain syndrome*, caused by a metabolic disorder which leads to a combination of brittle hair and an intellectual limitation in children born with this recessive defect. Finally, there is also the McKusick-Kaufmann syndrome, which is an extremely severe development disorder that was first documented in the case of the Amish because it appears more in this population group than in others.

Not Only for Humans

Even farmers know that they cannot keep breeding their animals for too many generations. When they start to notice a lot of disorders in their calves, piglets, lambs and foals, they realize they are dealing with the consequences of inbreeding. Inbreeding puts a halt to making animals better and more resilient. In Flemish, people will refer to this with the verb 'uitmendelen', which proves that Gregor Mendel's conclusions are well known in the world of farming. Actually the word is a nice and unexpected homage to the work of the Czech monk even though many farmers will not make the connection. But the farmer will know that keeping up this breeding process serves no purpose because genetics will protest against it.

11
Everyone's Genome Investigated

Technically and financially speaking, it is possible to generate a genome map of every newborn. But is that what we want? And how would we do that?

The high speed evolution of sequencing equipment and the plummeting costs of an entire genome sequence make this seem like a realistic scenario in the future. Each individual could be able to have their genome screened, in principle. And it is very possible to do this for every newborn.

In the year 2000, as a result of the International Human Genome Project, the first version of a sequenced human genome was presented. It took 15 years of hard work and more than 3 billion dollars (see Chapter 2). At that rate and cost, having an entire human genome analyzed would not be available for the general public, nor for research or medical use. However, quick technological advancements, a generation of new sequencing techniques (methods to determine the order of the 'DNA letters') and a strong sense of competition brought the market to believe that the *$1,000 genome* is a realistic goal. Strictly technically speaking, today it is possible to map out an individual's entire genome for this price.

An Ideal Help for Research and Diagnosis

For the medical world this offers very good prospects. Until now, geneticists have been obliged to make choices based on a diagnosis and clinical symptoms, and have only been able to look at one gene at a time. If multiple genes cause a disorder, the corresponding genes would have to be screened one by one. A labor-intensive, expensive and time-consuming procedure, without a guarantee for success. Some cases remained a mystery because it was not clear for many genes – or gene combinations – in what way they were responsible for causing a certain disorder.

Better and cheaper sequencing technologies, which map out the entire genome of an individual, offer hopeful prospects for genetic diagnosis and for better knowledge of a number of disorders.

One Step Further: Testing Healthy Individuals

Genetic testing is mainly used these days in clinical centers that study human genes in order to diagnose hereditary diseases. People who are tested often have a family history with hereditary disorders in different generations. In other cases testing is for patients who have symptoms that are not easy to diagnose, such as rare and often severe disorders. Meanwhile, a complete genome analysis is becoming more common in medical practice, mainly for disorders in which focused genetic analysis does not provide an answer (see Chapter 4). This immediately raises the question if it is possible

to use the same technology to discover any defects in healthy individuals. This would mean no longer testing to come up with a solution to the problem, but instead using a genome analysis as a method of prevention.

> **"The amount of generated data is so extensive that it is not easy to draw conclusions that are useful for a healthy individual.**

The philosophy behind this is that a greater knowledge of our genome makes it possible to map out health risks in healthy human beings. For these individuals, advice on how to live their lives and other forms of prevention could really improve their quality of life and enable them to stay healthy for a longer period of time. Besides, this technology could also be used to map disorders that we might pass on to our offspring, which could in turn help us make reproductive choices.

Widely Available on the Internet, but without Interpretation

Different companies offer individual consumers the opportunity to map out their entire (or part of their) genome through the Internet. Depending on the offer, prices may vary between $1,000 and $10,000. Intriguing detail: most companies will only provide the rough sequencing data and a limited analysis, but no medical interpretation of the genome. These laboratories will put you through to a physician for that part.

Yet the interpretation of the data forms the Achilles heel of this technology: many refer to this as 'the $1,000 genome and the $100,000 interpretation'. The amount of generated data is so extensive that it is not easy to draw conclusions that are useful for a healthy individual. There is no point of comparison, which is the case when interpreting a genome within a family that has known hereditary disorders. It is not possible to start the research with an index patient who suffers from a specific disorder in order to determine if an individual related to this patient also has a chance of developing this disorder.

The Interpretation of the Genetic Information Is of Utmost Importance

You can also ask yourself to what extent variations in the genome actually mean something. The human genome is very variable, with an average difference of around 1 base pair for every 1,000, between two randomly picked individuals. Most variations are clinically meaningless. Researchers must ask themselves every time if a certain variation is pathogenic (meaning it could lead to a disorder), possibly pathogenic or not at all. This is why they are constantly consulting databases that are built up internationally and where human genetic variations and their possible medical implications are brought together. This process has been mostly automated, but interpreting the work requires a lot of specialists.

The presence of a certain variation is not necessarily a guarantee of the development of a specific disorder. Many disorders have a lower penetrance (it is possible they do not even develop) and a variable expression (they can develop in different ways and at different times).

To illustrate this we can look at research James Lupski published in 2010. He is a famous geneticist who had his genome mapped to gain a better insight into a neurological disorder that affected him and his family, known as Charcot-Marie-Tooth disease. This is a disorder in which patients progressively lose muscle control due to malfunctioning nerves. Through analyzing his genome, he discovered that, aside from the defects causing his condition, there was also a mutation that should have led to a different disorder that should have killed him earlier in life. Yet this disorder was not bothering him, nor did it properly develop. What we can conclude from this is that there is information and variation in our genomes either that we interpret the wrong way or that gets bypassed by nature.

> **“The basic assumption for testing minors is that their own personal interest is put ahead of all others.**

There Are More Factors at Play Than Just the Genetic Ones

Moreover, a lot of disorders depend on a lot of different factors. This means that a lot of genetic variations play a small part in developing a disorder, and that these variations also are influenced by environmental factors. In Type II diabetes, for example, more than 65 different genetic variations were mapped out in the last few years. Individually they are only responsible for a small increase or decrease in the risk of developing this disease. This way they are a lot less predictive than other risk factors. Measuring the waistline, body weight and length, and looking at the family history can give a more accurate prediction of whether or not an individual will develop Type II diabetes than any other test.

Testing Newborns, Only If It Can Be Medically Justified

When we make the judgment to transform the same technology into an entire genome analysis for newborns, there is an even bigger challenge that arises. The basic assumption for testing minors is that their own personal interest is put ahead of all others. A possible medical advantage is a clearly justified motive to conduct a predictive genetic test. Let's think about the availability of therapy or preventive measures. If there are no well-founded medical reasons, predictive tests are usually postponed until the children have reached an age when they are mature enough, cognitively and psychologically, to make such a decision on their own.

Sequencing the genome of a healthy child is an infraction of the child's future autonomy and takes away the child's chance to make its own decision when the time comes. Children should be able to enjoy an open genetic future: they should make their own decision when they are old enough to do so. They should decide for themselves whether or not they want to undergo testing, know the test results and they should have a say in who else should know the test results.

The starting point of this debate should always be making as few choices for children as possible so children can make their own decision at a later stage in life. The only exception is when there is a clear medical importance in conducting the test, such as in the case of familial cancers that can make an appearance early in life, like retinoblastoma (a tumor of the retina) and specific forms of intestinal cancer. Even in those cases, sequencing the whole genome is not necessary. These principles are used in genetic centers, but they also apply in the context of screening.

“The past has shown us that the advantages of screening are often overrated.

Advantages of a General Genetic Screening Are Often Overrated

The past has shown us that the advantages of screening are often overrated. Research shows that people are over enthusiastic about the possibilities of screening and tend to forget about its shadow. People assume that early identification of a possible disorder can have only advantages. Estimating risk earlier in the process of life can only be a good thing, right?

With every screening it is important to look for the right balance between over- and under-diagnosing. Over-diagnosis will lead to unnecessary feelings of fear and insecurity, and eventually to unnecessary and possibly risky follow-up investigations and therapeutic procedures. Under-diagnosis will lead to an unjust reassurance.

The Example of Breast Cancer Screening

In order to illustrate the tension between under- and over-diagnosis, we refer to the example of breast cancer screening. To be clear, this is not a genetic screening but mammography in which an X-ray picture is taken of a woman's breasts. Some researchers believe that there is a clear health benefit; others believe that the disadvantages do not outnumber the marginal profits. On the one hand, its goal is to reach as many women as possible who have a chance of developing breast cancer. Yet a Dutch study showed that approximately 2 in every 1,000 women who have negative results after a breast cancer screening develop breast cancer before they do another screening. On the other hand, some women are falsely diagnosed with breast cancer. The same Dutch study calculated that 17 out of 1,000 breast cancer screenings are falsely diagnosed. Some of those women will undergo unnecessary treatment for tumors that will turn out to be benign. Still, the conclusion of the Dutch report was that population research for breast cancer is still worthwhile and does provide sufficient health benefits.

> **"It is essential to bear in mind a couple of core principles of screening.**

Whether in the case of children or adults, subjecting the genome to a thorough analysis without formulating a clear goal for the test beforehand provides all sorts of possible disadvantages that are inherent in the screening process: unrest about abnormalities, follow-up tests and treatment, unexpected complications, over-diagnosis, etc. That is why it is essential to bear in mind a couple of core principles of screening.

High Sensitivity and High Specificity

High sensitivity and high specificity of the test are of practical importance. High sensitivity means that the selection of genetic variations has to enable the researcher to distinguish between people who will develop the disorder and those who will not. A test with a high sensitivity detects the greatest number of people who will actually become sick and the number of false negatives is as small as possible. A high specificity means that the test will produce few false positive results. The number of people who get false alarms should be as low as possible. From all that has been mentioned about the genome in this book up until now, it should be clear that this is not easy when performing a genome analysis.

Health Benefit

A second core principle is the health benefit. What will the screening actually result in? To what extent will a certain test result lead to a greater quality of life or prevent an illness or death? To what extent does a certain test result help people to make informed decisions? Generally, if a test result does not offer clear health advantages, you should be asking why the test was done in the first place.

Well-Balanced and Easy-To-Understand Information

A third core principle is well-balanced and easy-to-understand information. This includes information about the disorder that is being screened for, the goal and reasons behind the screening, the test's reliability and its predictive value, and the possible advantages and disadvantages for the people involved and their family members. Based on this information, an individual that is offered a test should be able to decide whether or not to partake.

Cost Efficiency

The fourth core principle is how cost efficient the screening process is and if the cost is justifiable. What are the benefits of screening based on genome analysis compared to other screening methods when it comes to health profits or possible extra expenses (e.g. the cost of the test itself, follow-up research, etc.)? If the genome analysis leads to an unnecessary amount of follow-up research with no health benefit, it is probably not cost effective. Because of the government's limited means, public forms of screening should be carefully considered. If a complete genome analysis and the information that can be derived from it is relevant for public healthcare, in terms of actually being able to provide treatment or counselling, then it should be made a priority for the government. This is clearly not the case for a genetic test that determines the odds of a person becoming bald for example.

These Principles Also Go for the Private Sector

It would be wrong to conclude that these core principles are not relevant for the private sector. Forms of screening that do not meet these basic criteria can be financially disadvantageous for the healthcare system, even if the consumers pay for the initial tests themselves. It can push people to look for doctors to interpret the test results and can lead to further examinations and interventions.

A Complete Interpretation Is Still Music of the Future

Now that the genome can be mapped out at an acceptable price, everyone has the possibility to look at their own. This is undoubtedly a fascinating development. Yet we are still miles away from the use of a complete genome analysis as a preventive health screening method. Whoever faces the challenge of interpreting the massive amount of data from a complete genome analysis will have to deal with a lot of obstacles along the way. If you set out to explore your genome you will undoubtedly come home with an abundance of information of which the true meaning is unclear. Few doctors are prepared at this time to interpret the genome as a whole. Besides, this process requires a number of experts, from bioinformatics specialists to geneticists and medical experts. There is a chance that the genome analysis will lead to unnecessary other tests, unnecessary clinical interventions and unnecessary follow-ups. And that is aside from the possible psychological impact of these interventions. Analyzing the genome also has a lot of implications for your blood relatives. A discovered risk that can be of importance for your own health or that of your offspring can also be important for other relatives.

It is possible that in the future genome analysis will be used to screen healthy people for specific health risks. It will undoubtedly be about a limited set of genetic variations that have been carefully selected, are clearly interpretable, and will lead to health profit or reproductive choices, and will be affordable within our healthcare system. Turning the whole genome sequence inside out without a clear plan or intention might be a fascinating idea, but it does not necessarily lead to health profit.

12
Finding out More Than You Would Want to Know

Thanks to genome analysis, sequencing laboratories can trace more disorders. This comes in handy to confirm a diagnosis, but they also discover a lot of stuff that has nothing to do with the original question.

When diagnosing hereditary and congenital disorders, genetics has been playing an important role for many years now. Up until now laboratories have mostly focused on specific mutations to be able to confirm a diagnosis. Thanks to genome analysis we are able to trace disorders that are caused by one or a combination of genetic defects. It is a handy tool to fine-tune a diagnosis. But the wider researchers map out a patient's genome the more likely they are to find information that goes a lot further than what is required for the diagnosis.

One Gene, One Disorder or Many Genes, One Disorder

For a lot of hereditary disorders the culprit will be a single gene. To confirm the diagnosis, geneticists will look for that specific gene. This is the case for some muscular disorders or hereditary cancer syndromes (see Chapter 4). This approach works fine for disorders that can be explained by one or a limited number of genes. If there is a possibility that more genes are responsible for a disorder, existing technologies would examine gene by gene. This is a time-consuming and often frustrating approach. Thanks to genome analysis, which determines the sequence of the genetic code of all genes, it is possible to analyze different genes simultaneously. If this does not immediately give an explanation for a certain disorder, laboratory geneticists can make a wider search in the genome, with the risk of finding information that goes a lot further than what is required for the diagnosis.

Secondary Findings

Ever since genome analysis was implemented in clinical diagnostics, the question that has regularly surfaced is if laboratories have a professional responsibility to actively look for genetic variations that are outside the original clinical question. In 2013 the American College of Medical Genetics (ACMG) took a stance by saying that such a professional responsibility does exist and that a list of 56 genes has to be examined with every genome analysis.

According to the ACMG, these are genes that are known to cause disorders for which preventive or therapeutic interventions are available. They are mainly genes that play a part in familial cancer syndromes and congenital heart diseases. Based on the instructions of the American College the patient could not originally refuse the analysis of these 56 genes. If this directive was followed, this means that when a patient underwent genome analysis to find out the cause of, let's say, a congenital bone defect, they could also find out that they were prone to developing a hereditary form of intestinal cancer or a heart condition, such as hypertrophic cardiomyopathy (a thicker heart muscle, which can lead to cardiac arrest). Because these genes have nothing to do with the reason why the patient is undergoing genome analysis in most cases, we call them *secondary findings*. We can only ask ourselves how the patient, the parents in the case of a minor, or other relatives take the news.

Happy with This Extra Information, or Not?

Incidental findings are no exception in medicine. Think about a situation in which a patient undergoes a CT or MRI scan for a hernia and the doctors coincidentally discover a tumor that requires further examination. If the doctors were to keep such information to themselves, a lot of people would consider it negligent. We are dealing with medically relevant information that can lead to therapeutic intervention. In other words, it is in the patient's best interests to know about this information.

According to some, the comparison between secondary findings in the genome and the incidental findings in the scan is not a good one. With the scanning method, the doctor cannot ignore the tumor. In order to find out the information about the 56 genes, geneticists have to actively search for mutations in these genes, analyze them and interpret them. The information generated between CT or MRI scans and genetic tests is different: the scan gives an indication of a tumor that is already developing, whereas genetic testing just gives you information about the risk of whether or not you will develop a disorder.

The moment at which this disorder will possibly develop can still be a long way away. It is not certain to what extent it will happen, if ever.

> **“Incidental findings are no exception in medicine.**

Because of this, many healthcare professionals and researchers object to this approach where standard procedure requires extra genes to be analyzed – in this case the 56 genes that were chosen by the ACMG. Some find that this blurs the line between diagnostics and screening. If it would, in fact, be useful to examine these 56 genes, then it should be done with the entire population, and this approach would have to meet with the core principles of screening (see Chapter 11). Some also criticize the genes selected for the list: many of these genes have been studied intensively in families that suffer from their corresponding disorders. The chance is real that these mutations are less meaningful in individuals that do not have a family history for those disorders.

Some would, therefore, rather see the opposite: let's limit the chance of finding 'secondary' or incidental findings. This can be achieved by using 'filters' with genome analysis. Just looking at the genes that are possibly tied to the disorder and which lead the doctor to order the genome analysis will reveal less incidental information. Relevant 'lucky shots' can be communicated to the patient, but only if the patient is aware that these incidental findings could be a possible by-product of the test.

What Is the Patient's Role in All of This?

An important question is also what the patient's role is in this context. According to the approach of the 2013 American guideline, the selected 'secondary' results are part of the genome analysis basic package and therefore cannot be refused.

Others find that the patient's autonomy gives them the right to make their own decision when it comes to whether or not they want to receive the information. The main sticking point of this debate is the importance of incidental findings. In urgent, life-threatening situations, doctors will make decisions in the patient's best interests, even if they do not know the patient's wishes. Think about a situation in which someone is taken to the emergency room after an accident. Doctors will do everything in their power to keep the patient alive. Genetic information, on the contrary, in most situations is not 'urgent' information that is immediately life threatening. According to some, it is irresponsible to universally determine that a patient does not have a say in whether they receive secondary or incidental findings. In medical ethics it is generally accepted that a patient can refuse certain interventions, even if it is clear that this refusal might be disadvantageous.

The Right Not to Know

Some feel that when information that the patient has not requested is communicated to them that is an infraction of the right not to know. This right not to know has been included in many different legal systems and determines that healthcare providers need to respect the patient's wishes if they do not want to know about specific medical and genetic information. Article 10 (2) of the Convention for the Protection of Human Rights and Dignity of the Human Being with regard to the Application of Biology and Medicine: Convention on Human Rights and Biomedicine dictates: "Everyone is entitled to know any information collected about his or her health. However, the wishes of individuals not to be so informed shall be observed."

But this does not end the discussion. As with many other rights, the right not to know is not absolute. That is why there are a lot of exceptions in different European legislations, for example, when not sharing the information brings severe harm to the health of the patient or the health of others. Imagine someone who is suffering from a dangerous, contagious disease and chooses not to receive any information about their condition. Because of the risk of infecting others, the caregiver can and will have to inform the patient, despite their right not to know. It remains an open question if this exception should also apply to the genetic information discussed in this book.

The ACMG was not deaf to the criticism expressed of its policy and therefore updated its policy in April 2014 by giving patients the right not to receive this 'extra' information. This possibility has to be discussed with the patient before taking the test.

Genome Analysis of Minors Raises Even More Questions

Secondary findings raise some other questions as well. A lot of patients in need of this analysis are underage. This makes the situation especially complicated. The ACMG's opinion on this matter is that the patient's age does not make a fundamental difference. The idea behind it is that conducting a genome analysis is an opportunity that needs to be seized. It does not matter if the test is conducted at the age of 5 or 55. The effect of this approach could be that, for example, an 8-year-old girl, or in this case her parents, who undergoes a genome analysis because of an intellectual disability could find out that she has a mutation in one of the BRCA genes that could lead to breast and/or ovarian cancer, while this disorder will usually not appear when she is still a child.

Aside from whether or not there should be an active search for these genes, sharing this kind of information about minors means a complete shift of the current policy regarding genetic tests on minors.

The Right Not to Know in Practice

In 2008 the genome of Jim Watson was published in Nature (see Chapter 1). Was Jim Watson's entire DNA sequence published? No. Watson decided beforehand that he did not want to know if he had an ApoE4 variation. ApoE4 is the high risk variant of the ApoE-gene, the gene most associated with an increased risk for late-onset Alzheimer's disease. It is not a direct predictor of the disease - at least a third of patients with Alzheimer's disease don't have the ApoE4 variation and some that have the variation never develop the disease. Watson did not want to know if he carried the ApoE4 variation in his genome. Researchers had to respect his wishes and removed the data about the ApoE4 gene before they submitted his DNA sequence to the international DNA database.

Currently, policy recommendations would subject minors to predictive genetic testing only if it came with preventive or therapeutic interventions during childhood (see Chapter 5). Basically, they do not want to subject minors to these tests unless there is a good reason. We are dealing with decisions that are not urgent, and the children themselves could make the decision to undergo a test at a later stage in their lives. The simple fact that these minors, more than their parents, have to live with the results of the tests is a good reason on its own to let them decide whether or not they want to undergo them. Besides, you have to take into account the possible psychological and social impact of these tests when performed at a young age.

There Are Still Dilemmas

Such situations put healthcare professionals in the front line for a lot of ethical dilemmas. Delivering such news about possible risks, especially at such a young age, goes against the existing guidelines. By not informing the patient about the risk there might be some important information that is kept from the patient, which could also be seen as being ethically unacceptable. A possible solution could be to keep the information away from minors until they come of age. There have been suggestions to create systems in which different caregivers could consult the results of a genome analysis in all stages of life and use the information when necessary and relevant. To give you an example: it is not very useful to examine newborns for a disorder they might pass on to their own children. But this information might be relevant later in life.

A disadvantage of this approach is that relevant information might get lost. Besides, holding back information keeps causing difficult, possibly unsustainable situations for caregivers. The fact that such information about minors could also be relevant for parents brings some extra complications. If the parents are not aware of a possible hereditary risk of breast cancer this could lead to situations in which they pass this hereditary disorder onto other children, all while the healthcare professionals involved knew about the problem. The only way to avoid these situations as much as possible is to limit the number of secondary and incidental findings. The abovementioned strategy of 'filtering' leads to fewer dilemmas, although the last word in this discussion has not been said. Laboratories are currently looking for new standards to deal with this problem. At the same time it is important that different stakeholders, including the general public, think about this kind of issue. This is one of the reasons why this book was written and why a whole chapter was dedicated to this issue.

13
HELP! The Government Has My DNA!

Can biological samples be kept for research purposes? For how long? And under what conditions?

From many recent discussions it appears that it is often hard to find a good balance between enabling scientific research on the one hand and respecting the research participants' rights and interests on the other. One recent discussion that illustrates this problem is whether dried blood spot cards from newborns can be stored after they have been used for screening.

Newborn Screening and Storage of Dried Blood Spot Cards

In most Western countries newborns are pricked in the heel (or in the back of the hand) in order to catch some drops of blood. This is called the neonatal heel prick or Guthrie test. These drops of blood are then collected onto pre-printed collection cards often known as Guthrie cards. This normally occurs between the third and the fifth day after birth. The number of disorders that is being tested for using this procedure differs between countries. This way life-threatening but treatable disorders can be detected at a very early stage. The goal of neonatal screening is to prevent the development of certain disorders or treat the disorder at an early stage. Neonatal screening is widely accepted. Although this test is not mandatory in many countries, nearly all new parents choose to take it. After all, there is a clear health interest involved for the newborn.

But could these cards be used for purposes other than that for which they have been collected? And for how long? In the American state of Texas a court ruling determined that the long-term storage of these cards without the parents' consent is illegal. Because of that more than 5 million Guthrie cards were destroyed.

In 2012 the government of the American state of Minnesota decided to store the cards for no longer than 71 days. They were pressured into making this decision because of a threatening lawsuit. Similar debates happened in Western Australia, Ireland and Scotland.

Storing Them Temporarily Is Not a Problem

There are few objections to temporarily storing these Guthrie cards. Sometimes a certain analysis or test needs to be redone. In this case it is possible without taking a fresh blood sample. If an infant dies, the sample can help to determine the cause of death or to advise the parents on future pregnancies. Laboratories also use the cards to perform quality checks, improve existing tests and develop and fine-tune new tests.

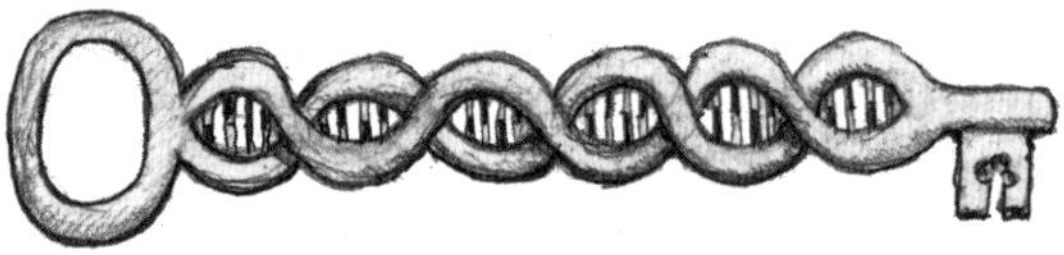

Longer Storage Could Also Have Its Advantages

For researchers, Guthrie cards are also meaningful research materials. In contrast to biobanks that collect biological materials from specific target audiences (for example, people suffering from a specific disorder), such a collection of samples gives researchers an oversight of a large group of newborns in a certain area. This enables researchers for example to analyze how frequently a certain genetic mutation or defect occurs in the population and in which regions. Or they can calculate the exposure to all sorts of toxic materials. When certain collections are stored for longer periods of time it is also possible to map the continuity of disorders, infections or environmental hazards. It is logical that researches seek to store Guthrie cards as long as possible to guarantee maximum usage of the samples.

From an international point of view the practices drastically differ. While Germany saves the samples for three months and Hungary and Poland only one year, countries like Denmark and Sweden save the samples for 1,000 years, so basically forever. In the Netherlands, the Dutch Institute for Public Health and Environment prescribes that the Guthrie cards should be saved for 5 years: 1 year for quality control and 4 more years for potential scientific research.

Advantages and Risks

The sensitivity lies in the fact that saving these Guthrie cards comes down to starting a DNA collection of the whole population. Theoretically, thanks to the technological advances in the field of DNA analysis and bio-informatics, storing this type of information enables the identification of people and the analysis of their DNA. Some fear that privacy regulations will not be respected, on the one hand, because of the fact that the blood samples would be connected to other medical and/or personal data, on the other because this data could end up in the wrong hands. Police and government services could use the data for goals beyond scientific research and medical purpose. Other observers are afraid that non-authorized people will be able to access this sensitive information.

The bigger the amount of data available for researchers the bigger the potential research advantages. Yet it remains important that we set out clear guidelines to determine the conditions under which the collection can be used in order to guarantee that the personal interests of the people involved are respected.

“It is logical that researchers seek to store Guthrie cards as long as possible to guarantee maximum usage of the samples.

Does Everyone Need to Give Permission?

The most important talking point in this discussion is which form of informed consent is most appropriate. According to a number of people, it is sufficient to notify individuals that research might be done. This is called the *opt-out model*. This model assumes that everyone wants to be part of scientific research except for the people who object to it. If parents have no objection to the use of the blood sample taken from their baby for research purposes, it will not be destroyed. Research could thus ensue depending on the approval of the project by an ethics committee.

Forensic Databases

A whole other discussion is the one about setting up DNA databases for forensic research. We are talking about the development of a tool that supports police services and justice departments with their fight against crime and to ensure public safety. But to what extent can the government use our DNA to protect us?

There are high expectations when it comes to DNA research being used by justice departments and police forces. Think about the identification of victims and perpetrators, tracking down suspects based on DNA traces, excluding innocent suspects, clearing up 'old', unsolved cases, or even profiling potential suspects.

Starting forensic DNA databases is only one possible technology for the justice department and police forces, next to many others like automatic recognition of license plates, screening phone calls, tracking data traffic, using cameras, a drone program, taking fingerprints, etc. For each of these technologies the question arises what the proper balance is between civil rights and liberties on the one hand and ensuring public safety on the other. One of the main tasks of governments is to protect their citizens, but at what cost? How many of our civil rights and liberties, how much of our independence, our privacy can be sacrificed for the *war on terror*?

When should someone's data be stored in a forensic database? One proposal was to do so for the entire population to be able to trace criminals more efficiently, the logic being that if you had done nothing wrong, you did not have anything to fear. As a recent example of this approach, Kuwait in 2015 passed a law making it mandatory for all its citizens and foreign residents to have their DNA entered onto a national database. But is storing everyone's DNA a proportionate measure? It makes everyone a potential suspect. When it comes to privacy issues, this measure gives us a run for our money. We cannot control the data that is collected and stored.

A similar discussion arose around the policy of the British National Forensic DNA Database. This database was founded in 1995 and contained over 6 million samples in 2012 (only 10% of which belonged to British citizens). Every month they collected approximately 30,000 samples. This database was the focus of discussion because the data of all suspects in all possible criminal cases was stored. In other words, once you were in, there was no way out. Not even if the case got dismissed or thrown out and not even when the suspect was found not guilty.

The European Court of Human Rights condemned the United Kingdom for this practice in 2008. In 2010 the UK changed its policy. If in the end you are not found guilty of a crime, ergo are not convicted, your data gets deleted from the database. If someone did their time behind bars, does the data still need to be saved? Different countries have different laws. In some countries, for example, only the samples of convicted sex felons are stored in the database forever.

We also need to make a distinction between DNA samples and the derived profiles. DNA samples can be collected from the suspect's body, but they can also be found at the crime scene. They are usually saliva, blood or semen samples. In order to be useable for identification purposes, forensic researchers will usually generate around a dozen short tandem repeats (STR) of DNA samples. These DNA sequences are found in areas of the genome far away from the genes. The variations found there do not translate into proteins and so do not have any medical purpose. This is important: it is not possible to use the STR in order to find out about someone's medical situation, as is the case with other forms of genetic testing. One of the STR traits is the repetition of a short DNA sequence (usually 3 to 5 bases). People who are not related have a different number of repetitions. By comparing different STRs it becomes possible to identify someone. The fact that the entire genome is not involved in this comparison gives some room for error and an unjust 'match'. This risk, albeit very small, is factored into lawsuits.

The difference between STR profiles and DNA samples is interesting from an ethical point of view, because DNA sampling is a sensitive issue. There cannot be an extraction of medical information from STR profiles, but, in theory, DNA samples can be reanalyzed at any given time and produce information. Therefore, some people oppose the long-term storage of DNA samples and ask to store only the profiles. The storage itself is not the problem here: the problem is the possibility that the information could be used for a different purpose. European countries have made different choices with regard to this matter. Some countries opt for the immediate destruction of the samples after obtaining the profile, whereas others allow for storing for a predetermined period of time, after which they are destroyed. Some countries even have different policies when it comes to the storage of data from a suspect or a convicted felon.

Critics object to this model because, according to them, people are not sufficiently aware of the fact that biological material (thus including DNA) is being saved. They would rather see a system involving "active permission" (an opt-in model), in which people declare explicitly that they agree with the fact that their biological samples will be stored for possible further examination for research purposes. Some countries, for example, will ask parents to sign the Guthrie card as a signal of their consent in the matter. Other critics think this is insufficient because there is no communication of the type of research that could possibly be conducted.

Samples from minors make the process extra complicated since the parents are giving their consent to (or are not resisting) the storage of the Guthrie cards of their children. If these children grow up to be adults, it is possible that they could object to the storage of their blood samples. There is also a need for procedures in this case to inform people about the storage itself and about the possibilities to object or to give their permission to use the cards.

In Reality, sometimes DNA Samples Are Stored

Sometimes it becomes clear in an unexpected way that DNA samples are being stored. In the case of soldiers, DNA is saved to identify the fallen. Guthrie cards can be used in a similar way to identify people. In the months following the attack on the twin towers in NYC, researches used such blood cards to identify the victims. After the firework tragedy of 2000 in Enschede, the Netherlands, in which 23 people died, there was also the suggestion of using these cards. Unexpectedly, there was public discontent about the fact that this collection of blood samples had been stored, which forced officials to figure out a different way to identify the bodies.

Studies show that people do not have many objections to this form of research in general. Yet in different situations it seemed that a lack of transparent communication about the storage and the use of such blood samples resulted in collective indignation and even the destruction of valuable research materials.

To clarify, in the regions where these materials were destroyed, the scientific research took place in full compliance with the local regulations and only after being approved by an ethics committee, but the population was not well informed on these matters. This shows that there is a need for transparency and good communication in order to safeguard trust in research.

Robot Photos Based on DNA

Recently researchers showed that it will become possible to make a profile of an individual's facial features based on a DNA sample – in other words 'a robot photo'. This research, on the intersection of medical imagery and genetics, offers tons of forensic, anthropological and medical applications.

These 3D images are developed based on a number of genes that influence facial features. This data is then processed with special software and then the computer will make a virtual reconstruction of the face. First experiences with this technology show that it works. You do have to be patient to see practical applications: the technology still needs to be further developed. More genes that impact on facial features need to be tracked and the virtual reconstruction needs to be improved. However, DNA alone is not everything. DNA does not tell you the age of an individual or what the lifestyle and diet patterns of that individual were.

This technology brought some fascinating reactions with it. Recently an artist used it to make virtual facial reconstructions based on cigarette butts and discarded chewing gum.

14
Genome Hackers

The internet is an ideal instrument for sharing genetic information between international research groups. But is our data sufficiently protected against hackers and cyber attacks?

Good Research Requires a Lot of Data

Through research we can gain more insight into the way that genetic and environmental factors like age, where a person lives, lifestyle choices, influence the development of certain disorders or traits. To make this research possible, it is essential that researchers can access large quantities of genetic data, information about disorders and health, and demographical data. This can lead to a better understanding of the role of genetic factors in the development of disorders, and consequently lead to better diagnostics and treatment.

To enable this research, there have been major investments into research infrastructures, like biobanks, in recent decades. These are collections of materials that contain DNA, like blood, urine and tissue, in combination with health and demographic data of patients and/or healthy volunteers. This information can be used for scientific research. There is often a distinction between population biobanks and clinical biobanks.

Known population biobanks can be found in Canada, the United Kingdom, Japan, Iceland, Germany and Estonia. These are usually large collections with materials from thousands of people. One of the goals of the UK Biobank, for example, is to gather the samples and data of 500,000 Brits between 40 and 69 years of age, and follow up with them for a period of 30 years. This way researchers hope to gain a better insight into the influence of diet patterns, lifestyle choices, environment and genetic factors on a person's health. When participating in this endeavor, volunteers are clearly told that their data and biological samples will be used for scientific research.

When signing a consent form, they are given an explanation about the general goals of the research and told that they can withdraw or change their minds about the use of their data at any time. Because of the research's long duration and the size of the data collection, researchers opted for a 'broad' consent, because they cannot predict all possible specific research applications beforehand.

Sensitive Information

A lot of hospitals started clinical biobanks, in which they store biomaterial of patients with certain disorders, together with data about their medical file. Sometimes it can be blood or urine, for example, which had been taken for clinical investigation. Often there is no need for explicit consent in order to store the samples, but patients can oppose their use for scientific research by notifying the doctor who treats them. This decision must not impact on their care. In other cases, the patient will be asked to donate some additional samples for scientific research. In these cases you do need explicit consent.

> **“The goal is not that our genome can be viewed, passed along, cultivated, analyzed or researched by everyone.**

Just like everything else in someone's medical file, genetic information is sensitive and confidential. Everyone's DNA is unique: a DNA sequence can identify someone. From DNA we can also extract potentially predictive medical data that tells something about the possible disorders that could develop in the future. Moreover, an individual's DNA can also be used to identify blood relatives and reveal medical information about those blood relatives. That's why genetic information needs to be treated with care. The goal is not for our genome to be viewed, passed along, analyzed or researched by everyone.

Privacy Goes before Scientific Importance

Researchers cannot just use data from patients or volunteers for their work whenever they feel like it. First of all, an ethics committee will investigate beforehand whether a research proposal meets all the legal requirements. In some cases the data will be made anonymous so that it becomes impossible to connect it with the identity of the patient or volunteer.

In other cases the data will be coded and someone other than the researcher will hold the key. Complete anonymity is often criticized because it becomes impossible to identify the participants in case research results would yield relevant clinical information.

The use of anonymity and coding is a powerful tool to protect one's privacy. Of course the unique character of a genetic sequence continues to worry many: it belongs to one specific individual, the only exception being identical twins.

Genetic Identification through a Detour

Is it actually possible to connect anonymous genetic sequences with an individual's identity? Until recently, scientists assumed that this would be extremely difficult or nearly impossible. In 2013 an American research team managed to reconnect anonymous DNA sequences to their original owner through the internet. The 'hacking' of the genome took place with some help from an unexpected corner. The researchers used genetic databases which are used to recreate family trees. First and foremost they distilled a number of short tandem repeats (STR) from anonymous genome

Anonymity and Fertility Treatments

Sperm and Egg Cell Donors: Not Always Anonymous

The fact that anonymity cannot be guaranteed with 100% certainty has implications for the donors of sperm and egg cells. A study showed that there are at least 40 companies that build a genetic database, using STR on the Y-chromosome, mitochondrial DNA or autosomal DNA (see Chapter 16). These three techniques make it possible to prove blood relations and identify the names of families that someone is related to.

This development, first and foremost, shows possible cases of non-paternity. It is possible to prove that the expected father is not the biological father. This can be because of adultery but also because of adoption, a change in last name or the use of a sperm donor in the context of fertility treatment. Years ago, there was a story that was circulating saying that one in ten fathers was not the real biological father. A recent study shows that in reality, it is (far) less than 1 in 100.

Thanks to the abovementioned genetic databases people can come into contact with close blood relatives they did not even know existed. If you just browse around the internet, you will easily come across success stories of people that were able to track down a biological parent, (half) sister, (half) brother or even more distant relatives. Also circulating are stories of donor children that managed to track down their biological father.

The abovementioned American study's goal was to find the identity of anonymous participants in a research project. Their data needed to be present in the consulted databases. If someone is looking for an anonymous sperm donor that contributed to his/her life, it is not necessary for the donor himself to be present in the database. It suffices that a close blood relative is in the database. This makes it possible to trace one or more last names. The combination of last name and possible unidentifiable information about a sperm donor that parents received through the fertility center could lead to cracking the code.

In a story that made the news, a donor child could trace his father using this method. He traced the family name through a genetic database. In the brief description the sperm donor had given while depositing his sperm in the fertility clinic, he mentioned that his dad used to play baseball and that he is in the American Baseball Hall of Fame. This combination of information made possible to identify the anonymous sperm donor.

Implications for Donors and Donor Children

It is unclear how many people voluntarily or involuntarily track down blood relatives. The few reports that appear online are success stories of people who were happy with the discovery and contact with 'new' family members. We do not know exactly how many people who actually found a close blood relative are happy with that and will reach out. This development raises three new questions for donor conception:

- Should we inform children about the donation?
- How do we protect the donor's anonymity?
- What are the dangers of unidentifiable information?

Anyone who has problems conceiving and resorts to fertility centers might need to get help from egg cell or sperm cell donors. In these situations, the parents are left with a difficult choice. Should I tell my child that we used someone else's biological materials to conceive? Is that really necessary? The 'social' parents might not be the kid's biological parents, but they take it upon themselves to provide the child with lifelong care. In the past people were advised to keep the child out of the loop, on the one hand out of fear of possible negative reactions to the use of this technology, on the other hand because of the conviction that this would only cause unnecessary restlessness. The social parents are the real parents, and a small amount of biological data does not change that. And how do you even tell this to a child? Is there ever a good moment to tell your children something like this?

Even though there are no exact numbers of people that tell their children about donors, it seems like nowadays there is a lot more openness than there was. According to some you are doing your child an injustice by not telling them. Studies with donor children show that only a minority would rather not have known about this information. They also clarify that children who heard the news at an older age are often shocked and get mad, while others are relieved. Different studies show that some parents struggled with the secrecy and feared that a family member might spill the beans at one point, so they chose to tell their child themselves. The fact that online testing could generate such information adds a small risk to the possibility of the child finding out that their parents had used donor material.

Sometimes the Legislator Intervenes

Countries like Austria, the Netherlands, Sweden, Switzerland and the United Kingdom have implemented legislation that makes it possible for donor children to find out the name of their donor when they become of age. This legislation sprouts from the view that children have a right to know their biological origins. In a lot of other countries, like Belgium, France, Portugal and Spain, sperm and egg cell donors are anonymous to the donor recipient, the defense of this point of view being that the donor could have acted for altruistic motives but is not interested in coming into contact with possible children born from their donation. This approach guarantees the absence of any financial, emotional or moral obligations. Purely pragmatically, some fear that the number of donors is going to drastically decline when anonymity cannot be guaranteed. Anonymity does not mean that donors are untraceable. The traceability of donors can be extremely important when there are hereditary disorders found in either the donor or the donor child. The new development of genetic tests for genealogical purposes is a possible deal breaker for legislation that is designed to protect the anonymity of sperm and egg cell donors.

Even systems based on anonymous donation provide future parents with some information about the donor. This information can be related to the appearance of the donor but also about the donor's line of work. Recently fertility centers have been trying to get donors to tell more about their motives behind donating. Although this information is in principle unidentifiable, as was mentioned above, the paradox lies in the unidentifiable information holding the key to actually identifying a donor, as was the case with the son of the famous baseball player.

The availability of online genealogical genetic testing is a new development that undermines the concept of anonymity even further. At the moment, the number of people who discover blood relatives by using these tools is fairly small, but no one knows how popular these databases will become in the future. It is clear that anonymity cannot be guaranteed 100%. This should get us thinking about how we communicate this privacy risk to donors in fertility clinics, about the possible impact this might have on legislation that protects anonymity and about the quantity of information we want to collect about 'anonymous' donors.

sequences that were at their disposal. An STR is a series of repetitions of a number of 'letters' or nucleotides in the DNA of an individual. They focused their research on the DNA of the Y-chromosome that is passed on from father to son. When the STRs were read, researchers entered them into two freely accessible genetic databases (www.ysearch.com and www.smgf.com). Together with their Y-chromosome, children usually inherit the last name of their father. They suspected that this method should make it relatively easy to track down a family name. They tested their hypothesis in a database that holds the genetic information of 911 people, in which they could check afterwards if the found last name was correct. In 12% of the cases they succeeded in retrieving the correct last name. The combination of the right last name in relation to the research data that was in the database (year of birth and American state) made it possible to determine whose DNA it was by simply consulting online phonebooks or online registries. In some cases they were able to bring the list down to one person; in others they ended up with a small number of candidates.

An Important Impact on Privacy

What does this mean for privacy? Does this mean the end? In any case, the scientific community was shocked by how easy it was to identify anonymous participants in a research project. Despite all the measures that were taken to protect the privacy of the patients and volunteers, it was still possible to identify them.

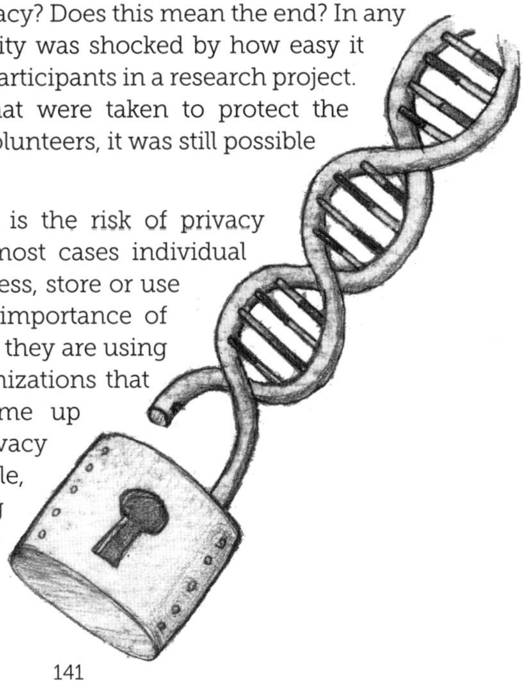

The question is: how great is the risk of privacy infractions in real life? In most cases individual researchers that collect, process, store or use data are well aware of the importance of protecting people's privacy if they are using their DNA. In the past, organizations that supported the research came up with strategies to prevent privacy infractions as much as possible, first and foremost by making the data anonymous, and second by rejecting data that could lead to identification, like date of birth. Besides, access to anonymous genetic sequences is becoming increasingly more the subject of supervision and researchers also have to give information on why they want to use the data. Researchers sign an agreement in which they promise not to share the data with others and not to use it in order to identify people. The data can often not be downloaded onto their personal computers, but needs to be processed through an online platform. The advantage of online systems is that it can be traced who accessed which data at what moment. In some cases the researchers cannot work with the actual data itself but a data manager will generate an answer to the research question formulated by the scientist. These are all measures to either avoid misuse of data or to make infraction traceable and therefore punishable.

It is also important to make a distinction between unauthorized access to a genome and the abuse of this information. It might be possible to make a connection between the DNA data and a specific individual; but the DNA sequence cannot always be interpreted immediately. It demands special expertise to analyze a genome and extract usable information. But no matter how small the risk is of identifying anonymous genetic data, it demonstrates that there is a need for legislation in order to protect individuals from possible negative consequences of intentional or unintentional use of private information. Many countries, for example, have laws that forbid employers or insurance companies from discrimination based on predictive genetic information, but regulations differ greatly from country to country.

15
A Neanderthal in All of Us?

We – modern people – have always been intrigued by this prehistoric relative. But are we really 'family'?

On January 2nd, 2014 *Nature* published an article that drew a lot of attention: the complete genome sequence of a Neanderthal from the Altai Mountains. In May 2010 there was already an article in *Science* called 'A Draft Sequence of the Neandertal Genome'; in other words, a draft version. The first time something was published about the genome DNA of the Neanderthal dates back to 2006, also in *Nature*, called 'Analysis of one million base pairs of Neanderthal DNA'. What does this tell us, except that scientists sometimes disagree on how to spell 'Neanderthal/Neandertal'?

Meet the Neanderthal

The Neanderthal (*Homo neanderthalensis*) lived in Europe and Western Asia. The oldest fossils of Neanderthals date back approximately 400,000 years, and the most recent ones about 30,000 years. In the meantime, the ancestor of modern man evolved in Africa, far away from the Neander valley, the river that gave the prehistoric human its name. The two human species evolved for a long time independently, which led to remarkable differences, especially when it comes to the skull and the skeleton – this we know because fossils love to talk about this, obviously. Why the Neanderthal became extinct is a question that still sparks many discussions.

> **“Chemically, DNA is extremely stable on the one condition that it is stored properly.**

DNA from a Neanderthal: Not That Easy to Harvest

To do genetic testing today lab technicians usually extract DNA from white blood cells.

Fortunately DNA holds a few special qualities. Chemically, DNA is extremely stable, on the one condition that it is stored properly. Deeply frozen, for example, in a solution that is enough of a base and, most importantly, not too acidic. Or in the case of a fossil, in soft chalky soil: this enables the DNA to be stored for thousands of years, usually in a piece of bone. Scientists used a piece of a Neanderthal's toe, a little under 3 cm, in the first genetic research into a prehistoric human. Researchers could estimate the age of the sample by estimating the age of the layer of soil from which it was excavated, approximately 50,300 years old, with a margin of error of 2,200 years, give or take. This small bone fragment held plenty of DNA to decode the entire genome of that Neanderthal in a reasonable fashion.

The First Research: 1 Million Base Pairs of the Neanderthal

In the article from 2006 the researchers talk about one million base pairs of a Neanderthal that lived about 38,000 years ago. One million base pairs – or DNA letters– is far from an entire genome. The human genome – and since the most recent article was published, also that of the Neanderthal– consists of more than 3 billion base pairs in total. The study that was done in 2006 only determined a fraction of the Neanderthal's genome sequence. This is for two reasons.

“The human genome – and since the most recent article was published, also that of the Neanderthal – consists of more than 3 billion base pairs in total.

First of all, the research was conducted with the first version of massively parallel DNA sequencers of what was then the latest generation (see Chapter 2; Jim Watson's genome was read using a similar device). This allowed scientists to read hundreds of thousands of DNA fragments, of tens to hundreds of base pairs, but at a cost that was approximately 10 times (or even more) more expensive than today. At that time it was still too expensive to conduct an entire genome analysis of the Neanderthal. Yet, it did allow the researchers to proudly announce that they had read one million unique pieces of DNA from the material that was extracted from the tiny piece of the Neanderthal's skeleton.

Secondly, many of the DNA fragments appeared to be from bacteria and mold that started to grow on the piece of bone that was used for the DNA analysis. Around 15% of the collected data was left out for this reason. Besides, the sequence of approximately 200,000 pieces of DNA did not match the corresponding sequences of the contemporary human and the chimpanzee, which pretty much made it incomparable.

It Was Well Done for Its Time

A good thing about this study was that they barely found any fragments that were a perfect match with the DNA of the modern man. If this had been the case, the researchers should have concluded that the sample had been 'contaminated' with contemporary human tissue: it is enough to just touch a small piece with your bare hand to leave behind loads more new DNA than the leftover original DNA that withstood the test of time. The research was a large undertaking: the researchers tested 70 pieces of Neanderthal bone and teeth before they came across one that still held measurable quantities of Neanderthal DNA. Most other bones either contained no DNA at all, or they contained DNA following contamination by archeologists and anthropologists who excavated the material and submitted it to further research. The right piece came out of the Vindija cave in Croatia and was excavated in 1980.

A Careful Interpretation of the First Results

After cleaning the DNA in the first research experiment, there were 15,701 DNA fragments left, of which the sequence resembled that of modern man and of the chimpanzee, but still showed a lot of variation. Researchers concluded –justly– that this was the Neanderthal's DNA. The pieces of DNA provided a total of a little over 750,000 unique letters, easily rounded up to 1 million for the title of the article. However, there was still one possible setback: the chemical damage to the old DNA. A letter was sometimes replaced by another one (Cytosine (the C from AGTC) changed to Urasil, or Guanine (the letter G) changed to Xanthine, for those of you out there who care about the details). The interpretation was not to be taken lightly.

"The one million dollar question: did the Neanderthal and modern day man cross paths again in more recent times and did they exchange genetic material?

By counting and studying the differences between these DNA fragments and the corresponding reference sequence of the modern man, researchers concluded that both species were related and had a shared history. The collected data sufficed to prove that the Neanderthal's ancestors separated about half-a-million-years ago from the ancestors of the modern man. They calculated this by comparing the differences between modern day man and apes, between modern day man and the Neanderthal, and between the Neanderthal and the apes. It shows that 8% of variation had appeared in the human genome since the Neanderthal separated from modern man's ancestor. On the scale of 6.5 million years of evolution from ape to man, this is almost 500,000 years.

Genetic Exchange between Neanderthal and Modern Man?

The one million dollar question: did the Neanderthal and modern day man cross paths again in more recent times and did they exchange genetic material? In the first investigation scientists carefully suggested that this was possibly the case based on some specific genetic differences and similarities. But they also suggested holding off any declarations of that sort until they had more complete DNA results from other fossil remains of Neanderthals.

The Second Research Paints a Better Picture

The results and calculations from later publications show that nothing is certain until there is enough data: the researchers came to the conclusion that 10 to 40% of the sequences that were used in the first study must have come from more recent human tissue. This does not undermine the first researchers; it just shows that comparing differences and similarities is extremely difficult. So they improved the technique, installed a clean room for DNA isolation and one for DNA sequencing, and they aimed to report 3 billion base pairs in the Neanderthal. Even more bone fragments from the Vindija cave were analyzed, from different Neanderthals. This enabled researchers to paint a better picture of the individual differences and the mistakes that found their way in the sequences through chemical reactions. With all the numbers and counts, they now concluded that Neanderthals and modern man split ways 825,000 years ago.

Clear Differences between Apes, Neanderthal and the Modern Man

The fun part of this study is that enough information was available to analyze a number of genes of the Neanderthal's genome and to compare these to corresponding ape genes and human genes. For a number of genes it could be predicted that the protein they produced would have different effects. An example can be found in a gene that could play a part in the pigmentation of the skin, and which slightly differs between man, ape and Neanderthal. Other examples include certain genes that play a part in intelligence, cognitive behavior, etc. A lot of these differences still need to be examined, but this study draws a picture of how genetics can explain the differences between humans and primates based on what appear to be miniscule genetic differences.

Better Neanderthal DNA from Altai in Siberia

In 2008 researchers discovered a Neanderthal fossil in a cave in the Altai Mountains in Siberia: a piece of a Neanderthal's finger. DNA sequencing showed that it was a female finger. The circumstances of discovery, storage and research were ideal, so lab technicians in 2013 were able to determine the entire genome of this Neanderthal thanks to this fossil. The quantity of 'modern' DNA was estimated at 1%, which proved that contamination with modern DNA was out of the question. On the contrary, there is a convincing similarity with DNA results from previous studies, which confirm that this is in fact a piece of the Neanderthal's finger.

From the new research results, scientists concluded that the ancestors of modern man and the Neanderthal parted ways somewhere between 553,000 and 589,000 years ago. This confirms the earlier estimations.

The Genetics of Neanderthals Provide Insight into That of the Modern Man

Why is further research into the Neanderthal's genome useful and needed? The last shared ancestor of modern man and the Neanderthal disappeared half-a-million years ago. The answer is that the differences in the genetic composition of Neanderthals and modern man give us information about how man evolved since then – and which genes played crucial parts in that evolution. Because in those last half a million years man evolved to what he is today and the way he looks: evolution to you, me, and all other people.

The RUNX2-Gene: At the Origin of the Difference between Neanderthals and Modern Man?

People who are born with a defect in the RUNX2-gene nowadays have distinctive skeleton and teeth defects. The most noticeable characteristic is the partial or complete absence of the collarbone. This rare disorder, with a one in a million frequency is called cleidocranial dysplasy. An important difference between modern man and early man-like creatures, including the Neanderthal, is the way the skull is built. Based on this, researchers formed the plausible hypothesis that with this difference in the function of the RUNX2-gene, modern man strayed from the Neanderthal's path. This is probably an oversimplification, but the example gives a hint on how species evolved.

Another Human Species: Denisovans

Also in 2008, and again in the Altai Mountains, in the Denisova cave, archeologists found fossil remains that they suspected came from a Neanderthal. After DNA analysis from the piece of finger that they found, it appeared that the fossil belonged to a different prehistoric human species, the Denisovans (*Homo denisova*). This human species lived in Eurasia in the same time period during which modern man made his appearance in Africa. Research showed that the last shared ancestors of the Denisovans and Neanderthals lived 445,000 to 473,000 years ago. It appears that this species evolved from a different migration wave of man-like creatures coming from Africa than the ancestors of Neanderthals and modern man.

DNA Research in More Recent Ancestors

A completely different story about the genetic profile of our ancestors is that of Spanish hunters who lived in a more recent episode in the history of our origins. In 2006, in a cave in Northern Spain, archeologists found the skeletons of males that lived approximately 7,000 years ago. They were not Neanderthals but ancestors of modern man still living in that region.

The genomes of these men were researched. This showed that there was a strong relation to Scandinavians and British people, which led to the provisional, cautious conclusion that all European hunters came from the same northern European descent.

The published data also contains specific genetic information about the habits of these hunter-gatherers. They only had a limited number of copies of the amylase gene, which is responsible for a saliva enzyme that converts starches into sugars.

Modern Europeans mostly have eight or more copies of that gene. This led researchers to believe that these Iberian brothers could not digest milk and probably struggled with a diet that was rich in starch.

Another remarkable discovery is that early Spanish people had a 60 to 80% chance of having blue eyes. There are strong genetic indications that their skin was heavily pigmented: apparently fair skin is a recent development in modern man. These two early Europeans were also not malaria resistant. The genetic information showed that their immune system was clearly focused on other diseases. But it is hard to make firm conclusions that apply to an entire population based on the study of the genome of these two individuals.

Even More Genetic Exchanges between Prehistoric Human Species

Genome research shows that the Neanderthal has more genetic variations in common with people currently living in Eurasia than with the current inhabitants of central Africa. This tells us that there was in fact an exchange of DNA in the northern regions between Neanderthals and the first modern humans when they were swarming out of Africa to settle in the rest of the world. This mix of DNA probably helped modern man more easily adjust to the colder climate. There are also indications that several disease mutations, such as mutations related to Type II diabetes, originate from

Neanderthals. Europeans and most Asians possess approximately 1.5 to 3% of genetic Neanderthal material. Other Asians and the indigenous inhabitants of America share even more genetic material with Neanderthals, probably due to an extra exchange. But also the Denisovans left their genetic information with the current population of Oceania. And there, new discoveries do not seem to stop: the Denisovans appear to possess an amount of genetic material that comes from another, more unknown, more distant ancestor that must have lived somewhere between 1.1 and 4 million years ago. It might be the *Homo erectus*, but the odds are incredibly small that DNA can ever be recovered from the fossil remains of this human species.

A Puzzle That Sketches Human Migration

Even this data is just one piece of the puzzle in the study of migration and population, evolution and interaction of modern man. Each new historical genome sequence will, combined with archeology and other findings, complete the puzzle further. This chapter only shows a fraction of the research results, an impression of what is known today about the history of human origins. Further genetic research of DNA from fossil materials will without a doubt add to knowledge of human prehistory.

16
Mudbloods

Racial purity earns a place in today's popular culture because of Harry Potter's 'mudbloods'. To what extent does genetics contribute to the facts or prejudices that exist about this topic?

In J.K. Rowling's books the term 'mudblood' has a highly offensive meaning. In the eyes of the 'real' wizards it means a wizard or witch that comes from 'dirty blood', as explained in *Harry Potter and the Chamber of Secrets*. Mudbloods are considered to be of lesser value because their parents are not wizards. And 'squib' is the curse word for someone that has wizard parents but cannot do magic themselves. Racial purity obviously occupies mankind. Unfortunately there are examples in which genetics is used to test and promote racial purity.

The Call for Racial Purity Is a Controversial One

In 2012 there was a commotion when it came out that a Hungarian member of parliament and member of the extreme right-wing Jobbik party had his DNA tested by a commercial company to find out if he descended from Roma gypsies or had Jewish roots.

> **“The further you go back in your family tree, the greater the odds are that you will find the last name of your partner.**

That is possible because in every population group a number of specific genetic variations appear. The test result, according to the company, showed that the politician did not possess Jewish or Roma variations. The story went public when an extremist website praised the politician's approach and preached racial purity. This brings back unpleasant memories of the period before and during World War II when races were separated and the Arian race was granted superiority. In Hungary this led to charges being pressed with the court and national and international moral indignation grew. Not only does a scientific basis for genetic purity not exist, it is also ethically despicable to want to prove this scientifically.

More Than One Billion Ancestors

We all have two parents, who each also have two parents (so 4 'ancestors' in total), who each also have two parents (8 in total). If you go back seven generations we get to 128 ancestors, make that 20 and you get to 1,048,576 ancestors, and if you top it up to 30 you get over a billion. Of course these ancestors are not all unique ancestors. This theoretical conclusion does not consider the fact that some parents married within the family tree. This calculation does, however, get the point across that the genetic variation found in an individual is an extraction of a lot of variation in a lot of ancestors.

> **"People individually differ in a couple of million letters in their genome. Between all these divergences there are the so-called private variations, rare variations that only appear in families or individuals that originate from the same region.**

Do you know the family names of your 8 great grandparents, or the names of your 128 great-great-great-great-great grandparents? The further you go back in your family tree, the greater the odds are that you will find the last name of your partner. It is clear that all these ancestors helped shape our genetic constitution. But it is impossible to determine our real origin: we inherited genetic material from all of our ancestors and that gives us many 'origins'.

Abraham and His 46 ChromoSONS

A first technique to track your ancestors is through the Y chromosome. Only males have a Y chromosome and they always pass it onto their sons (otherwise they would be their daughters). Because males only have one Y chromosome, there is no exchange of genetic information like there is between other chromosomes (except for what is called the "pseudoautosomal regions" of the

X and Y chromosomes, but that is a technical detail). In other words, the Y chromosome is passed generation after generation unchanged. There are a couple of exceptions: it happens that copy errors appear when the Y chromosome is passed from father to son. Such differences separate a population's Y chromosomes from another's. Researches can use this as the basis to connect different Y chromosomes from the same origin and go back centuries. Worldwide this method recognizes different basic types (formally known as haplotypes) and can trace distant relatives in the paternal line.

Local Women

A second technique uses the mitochondrial DNA (mtDNA). A small quantity of a human's DNA is not in the 23 pairs of chromosomes of the cell's nucleus, but outside that, in the mitochondria. This mtDNA is generally only passed down from the mother to her children, and this is over generations.

This makes it possible to determine who is related to whom on the mother's side. A DNA analysis from the paternal side as well as the maternal side can deliver information about which genetic group someone belongs to. Then it is about mitochondrial haplotypes. For instance, in a study about the African Lemba tribe, comparison of the mitochondrial DNA showed that there were no similarities with haplotypes that frequently appear in the Middle East. This means that male ancestors who migrated to Zimbabwe married local women. This is a fine example of how population genetics can help us better understand history.

Research from the mtDNA shows that all modern people originate from Africa. This is where the story of 'mitochondrial Eve' comes in – a term first used in a publication by genetic scientist Rebecca

Mitochondrial DNA

Mitochondria are remains of an old cooperation. They originated from the evolution of the earliest higher animal species and its bacterial ancestor. We had then just reached the single-cell organism stage. A couple of other early life forms, ancestors of bacteria, had developed a system for using oxygen, and our single-celled ancestors did not. It was an extraordinary change for the early ancestor to integrate these bacteria in its system and to live out the rest of time in symbiosis. That is the reason why mitochondria have their own DNA that strongly resembles that of bacteria.

Mitochondria are the cells' energy factories. The mitochondrial DNA (mtDNA or mDNA) consists of 37 genes, of which 22 play a part in the conversion of oxygen into cell fuel. mtDNA is interesting because it is not located in the nucleus, and does not matter when it comes to sexual procreation. Imagine the size of a sperm cell or egg cell: the sperm cell is small; as a matter of fact it is just a cell nucleus draped in a little fuel to cover the trip to fertilization. The egg cell is large (sometimes exceptionally large, like in chickens, for example), and provides all the nutrition that the early embryo needs and for the cellular compartments, including mitochondria. The female gets the furniture, provisions and half the DNA quantity for the cell nucleus, whereas the man only provides the other half of the DNA (which the fertilized egg cell cannot do without).

Cann and Allan Wilson in 1987, so long before Bryan Sykes wrote his book *The Seven Daughters of Eve* about the same topic.

Moreover, mtDNA research was also used to identify the Russian Tsar Nicolas II and his children, who were maltreated with sulfuric acid and murdered in Jekaterinburg in 1918. The Tsar's wife, Alexandra, and his children Olga, Tatiana, Maria, Anastasia and the young Tsarevich Alexei were related to Queen Victoria on the maternal side. With the help of DNA from living relatives, researchers could confirm that the remains found in the (mass) grave actually belong to the Tsar and his family.

> **“Regionally people look more alike than they suspect because of the borders that were drawn for political, historical or religious reasons.**

Similarities and Differences

A third technique uses the DNA of other chromosomes ('autosomal' DNA). We know that the DNA between two individuals is 99.9% similar. This technology focuses on how much that last 0.1% makes us different. Based on the research between the similarities and differences of thousands of individual genetic variations it is possible to determine statistically if someone has more or fewer genetic variations in common with a specific population group. Or, in other words, to see how far someone is removed from someone else in the family tree. When researchers compare the genetic variations of different individuals in different regions or countries, the individuals that originate from the same region will 'cluster'.

The Illusion of Racial Purity

Individually people differ in a couple of million letters in their genome (see Chapter 3). Between all these variations there are the so-called private variations, rare variations that only appear in families or individuals who originate from the same region. This is possible because a variation can happen all of a sudden and therefore is passed on only locally between generations. There are a number of published scientific studies that use such unique variations in order to trace an individual's origins. And that seems to work surprisingly well.

For example, in a particular study researchers could access the genomic data of hundreds of patients with anorexia and a control group from 13 European countries plus the United States and Canada. The goal of the original study was to search for genetic variations related to anorexia. But a by-product of the study – in any case a piece of information that made the interpretation of the results a lot more difficult – was the observation that the different participants could be separated based on genetic variations that linked them to their origins. In other words, almost everyone that was originally from any of the participating countries could be placed back on the map in their respective countries, so to speak. To top it off, even the neighboring population groups had more variations in common. The researchers could place the participants on the map, north to south, east to west.

The researchers could even measure the distance between people, based on the statistical analysis of the frequency of the hundreds of different genetic variations. An interesting fact is that the Finnish, on average, are further removed from surrounding population groups than, for example, the central European population groups, or the people from the areas around the Mediterranean. This probably does not seem very odd to someone who knows a little bit about European history and geography. What obviously does not work is placing the Americans and Canadians neatly on a map of North America in the specific area where they live. And this is not surprising, because most of them have European roots. In any case, these kinds of studies paint some nice pictures, and show that the studies for genetic determinants of anorexia and other conditions become harder when patients and controls from different origins are mixed. Moreover, it shows that it is possible to trace someone's origins by using these so-called regional or private variations.

Various companies offer genetic tests in which they will determine if you show more comparisons with variations that are typically 'labeled' African, European or Asian. Of course, the quality of the comparison becomes stronger when the reference population has been mapped out well. An example of this can be found in the

Population Genetics: Differences between the Belgians and the Dutch

Luigi Luca Cavalli-Sforza is an Italian population geneticist. He studied the genetic variations and connected that information to historical, cultural, geographical and other surrounding factors. He was born in Genoa in 1922 and headed to the United States where he would become a professor at Stanford University in 1970. In 1993 Cavalli-Sforza published a study in which for the first time ever a series of genetic variations of 26 different European population groups were compared. It was about simple variations because in 1993 we were still at the beginning of the 'genome era'. It is interesting, however, that he showed that genetically Belgians are closer to Austrians, Germans, and the Swiss than to their upper and lower neighbors, the French and the Dutch. The explanation for this fact is clearly historical: the territory that is now Belgium was occupied for a long time by the Habsburgs and apparently these were relatively 'fertile' times in comparison to the other periods when the French, the Spaniards and the Dutch ruled. The Dutch are closer to the Danish, the Belgians are somewhere in the middle of the Swiss, the Germans and the Austrians. Altogether we are furthest removed from the Lapps and from the Sardinians, two population groups of which it is known they have other origins. Besides that, we see a relatively nice correlation from north to south and from east to west. It will not surprise anyone the Swedes 'cluster' with Norwegians, Portuguese and Spanish with Italians, Hungarians with the Polish and the Scottish with the Irish.

Today, thanks to complete genome analysis population groups can be examined in greater detail and more local heterogeneity can be found. Unfortunately, no other studies comparing Dutch to Belgian individuals have since appeared. This makes it hard to determine if Cavalli-Sforza's findings are true. But there is also no reason to doubt them.

Other studies have shown that, apart from mountains or other natural obstacles, the spoken language and religion form strong break lines in genetic variation. Also, Europe has been the stage for thousands of years of migration and hundreds of years of war and other disasters, and this led to the variable mix of genes and genomes.

story of Vy Higgensen. In 2007 this African-American lady took part in a TV show to trace her roots. From similar testing conducted by three different companies, it appeared that her roots could be traced to three different African tribes: the Mende tribe near Sierra Leone, the Wobe tribe in the Ivory Coast and the Mendenka tribe in Senegal. This illustrates that there are four problems when trying to determine someone's origins. Firstly, it is not possible to tie one's origins to one specific place. It is an illusion to believe that genetic testing can do this. Because of our widespread number of ancestors, our history is by definition linked to a history of histories of our ancestors. Secondly, this anecdote makes it obvious that different companies use different genetic variations. Some use more variations of a certain region and can therefore produce more accurate results with regard to that specific region. Other companies do not even cover some areas or just include a small number of variations in their tests. These differences greatly impact on the outcome of the test. Thirdly, we also need to consider the fact that the current population groups are not perfect representatives of ancient populations, but only represent a 'genetic proxy' of a certain region, group or tribe. Fourthly, these companies each use different ways to express the percentage of similarity a certain individual has to a specific population group.

Identity: Genetics or Politics?

Genetic analysis is an interesting instrument for contemporary scientific research into population genetics. While clinical genetics mainly focuses on family-related diseases, population genetics maps out populations from a historical, as well as a contemporary, perspective. But how dangerous is it exactly to connect genetic technologies with an attempt to construct a political-, religious-, ethnic- or family-based identity? Attempts to connect race and identity recall dark images from mankind's history. It makes us think

Vikings in Sicily

In 1996 Leuven-based researchers found mutations in the PMM2 gene, responsible for a very rare, congenital metabolism disorder, a 'congenital disorder of glycosylation' (CDG). After this publication was released requests for genetic tests came in from all over Europe. Two CDG patients from Sicily, brother and sister, appeared to also show mutations in the gene, which confirmed the diagnosis. One of the mutations was named the 'Viking mutation' because it was mainly seen in patients from Denmark and the southern part of Sweden. The Sicilian doctor who got a copy of the mutation report on his patients sent researchers a map of Europe that showed the travel paths of the Vikings: and yes, the Vikings went all the way to Sicily. They clearly left some genetic material there. Does this make the brother and sister Vikings? Not quite: mixing genetic information with that of the local population dates from too long ago. But they do carry traces of the turbulent history of their island in their genome.

The 'age' of the 'Viking mutation' in PMM2 was estimated to be around 1,500 years. This is calculated based on the number of generations needed to account for the mutations on the same chromosome in current patients. This might correspond with a kingdom that existed between the 8th and 10th centuries and was made up of Denmark and the southern regions of Sweden. Various seamen made great sea trips in that period. From Normandy, their offspring made it all the way to Sicily in the 11th and 12th centuries. It is not clear if the Vikings themselves already visited Sicily earlier. Some historians claim that they reached Constantinople and the Black Sea in the 9th century, thus passing Italy's southern region.

CDG-PMM2 is a recessive disorder. Carriers usually do not show any symptoms and can easily 'take it with them' on their adventure quests.

of the segregation between blacks and whites in the southern part of the United States in the first half of the twentieth century. Separate toilets, separate bus seats, train seats, restaurant seats, theatre seats, separate sports games, in short separate everything. In some states, blacks and whites were officially separated by the so-called *one-drop rule*. If one of the four, eight or sixteen ancestors was 'black' –the criteria changed from state to state– the person involved was immediately 'declassed' and received a lower social status.

Even today the determination of one's ancestors plays a part in receiving certain advantages, for example in rules about descent to determine who is a member of an Indian tribe. These were introduced to determine who can enjoy the financial advantages of the American government or who can actually say that he/she is a tribe member. Whoever believes that a genetic test is capable of

GoNL: 'The Dutch Genome'

In the Netherlands there is a project running that wants to analyze the genomes of 250 trios (i.e. the affected child and their two parents). The participants were selected from different regions in the country. From both a scientific and a clinical perspective it is important to chart the 'normal' variations in a population group. This allows researchers to better distinguish rare variations, which only appear locally but are clinically not harmful, and variations that might be associated with certain disorders. Besides that, the data is also useful for population genetics on its own but also for forensic research. The Netherlands has great expertise when it comes to biobanks and this initiative serves that purpose. The list of variations has already been published and also appears to be useful for genome research with Belgian patients. The Belgians and the Dutch are not that different, no matter what Cavalli-Sforza's findings may state.

In Belgium there is no similar project on that great scale. Since 2013, however, there has been a project called the *Belgian Medical Genomics Initiative* (BeMGI), that does want to collect all the genomic data that has been documented in a clinical context in the different genetic centers and could prove useful in documenting variations. This could be genetic information on individuals that were checked in studies or parents of children with a clinical disorder, for example.

determining who really belongs to an Indian tribe, or who is really Jewish, or belongs to a certain class of people, is confusing genetic identity with political identity. Mankind's history is a process of migrations and mixing population groups. Stating that genetic purity is a goal is a misplaced political idea that is built on genetic quicksand. From a genetic point of view the concept of race does not even exist. Sometimes the genetic differences amongst the same population group are greater than some differences between different population groups. That is why, for example, some Arabic people have, statistically speaking, more genetic similarities with the Jewish than with the average Arabic person. Biologically speaking, genetic purity is pure nonsense.

17
My Olympic DNA

Being good at sports is genetically determined. Compare the participants in a sprint race with those who compete in weightlifting.

Your genetic makeup influences, for a large part, if you will be good at sports and even which disciplines. Just think back to gym class and you can remember the pupils that excelled and the ones who did not. Heredity influences your height, the length of your arms and legs, the size of your heart and other muscles, the volume of your lungs and the flexibility of your joints. Genes partially determine how your body reacts to training, diet and other environmental factors. These elements greatly influence if you should prepare for a career as professional athlete.

Genetics Determine – for a Large Part – Your Athletic Performances

It is no coincidence that a lot of professional basketball players are taller than 2 meters, top swimmers have broad shoulders and a slim waist, the best cyclists have a large lung capacity, mountain climbers have a large number of slow fibers in their muscles and sprinters have a large number of fast fibers.

Whoever lacks these qualities or possesses them on a smaller scale will have to score exceptionally well on other qualities like technique, tactical insight or mental strength to be able to get to the top.

“Your genetic makeup determines for a large part if you will be good at sports and even which disciplines.

Genes for Endurance and Genes for Speed

Researchers have tried to identify the genes that influence athletic performance. In total there are over 200 genetic variations that possibly play a part in athletic performance. This describes the genes that have an effect on the functioning of the heart and breathing, next to the genes that influence the muscle functioning and blood circulation. Two genes in particular get a lot of attention, namely the ACE gene and the ACTN3 gene.

A 1996 study performed on some members of the Australian rowing team showed how variations on the ACE gene may influence endurance. They found the genetic codes for an enzyme that, amongst other things, tightens the blood vessels and stimulates the growth of skeletal muscles. The study showed that the strongest rowers in Australia often had an I variation of the gene. This variation makes the gene a bit longer causes the enzyme to be less active so that the blood vessels are tightened less when the muscle is under pressure. The muscle works more efficiently this way and has more endurance. This variation in the ACE gene is also often found with mountain climbers.

The ACE gene also has a D variation, which lacks the 'weird' DNA fragment and makes the gene shorter and enables it to perform better. This variation is tied to short powerful endeavors. In a study conducted on top swimmers, the D variation was found more frequently in short distance than long distance swimmers. The long distance swimmers were more likely to have the I variation.

In other studies, the I variation was statistically tied to athletes who focused on endurance, whereas the D variation was more present in athletes who rely on explosiveness and strength. But this is not an 'all or nothing' effect, and this finding could not be confirmed by all studies.

Exceptional Genes

In exceptional cases, particular genes can have a remarkable impact on athletic performance. There is the case of the Finnish skier Eero Antero Mäntyranta (1937-2013) who won a total of seven medals in the Olympic Winter Games in Squaw Valley (1960), Innsbruck (1964) and Grenoble (1968). He always got those medals in the long distance events like the 15k, 30k or 4x10k. During his career he was suspected of doping because his red blood cell count was 25% higher than average. This allowed his blood to take in more oxygen and enabled him to ski longer and at higher speeds. In 1993, thirty years after his amazing performances, researchers showed that Mäntyranta, just like other members of his family, had a 'mutation' or 'defect' in the EPOR gene (the receptor for EPO or erythropoietin). This mutation was responsible for the high production of red blood cells. Despite a training schedule much like those of his fellow professional athletes, he had a comparative advantage when it came to endurance. Because of the defect he was also diagnosed with polycythemia, a disorder that characterizes itself by headaches, high blood pressure, heart conditions and a higher chance of a stroke. Eero Antero Mäntyranta died in 2013 from a heart attack.

Some gymnasts and dancers have a hyper mobility that can be explained by disorders in the connective tissue, like Ehlers-Danlos syndrome (type 3), Marfan syndrome or mild forms of *osteogenesis imperfecta*. Miranda Maldonado, for example, is a young professional gymnast who suffers from Ehlers-Danlos syndrome. This makes her joints abnormally flexible. In the long run this can lead to instability of the shoulders, knees and other joints. Often these athletes become patients who have to deal with chronic pain at a young age.

The ACTN3 gene impacts on the proportions of two different types of muscular fibers in terms of power exertion and endurance. Type 1 muscle tissue has a slower and less powerful contraction, but it can endure the contraction longer than type 2. Type 2 muscle tissue has a contraction that is more powerful, but cannot endure it as long as type 1. A larger quantity of type 2 muscle tissue is more useful for sprinting and power training, whereas a larger quantity of type 1 muscle tissue is favorable for endurance-related disciplines. A 2003 study analyzed these genetic variations in 429 professional athletes of Caucasian descent in 14 different disciplines and compared these to a control group consisting of 426 Caucasian individuals. This study showed us yet again the possible role of the ACTN3 gene in the development of the proportions of the type of muscle fibers. Other studies taught us how different variations had a different impact. The 577XX variation is tied to the absence of the alpha actinin 3 protein production and leads to the creation of more type 1 muscle fibers, which has a positive influence on endurance capability. The 577RR variation is associated with the normal production of this protein, which allows the muscle to contract faster and more powerfully, which is an advantage for sprinting or other explosive activities. Studies with professional football (soccer) players showed that the majority had the 577RR variation, where the majority of rowers had the 577XX variation. Other studies failed to confirm this connection. It seems a little bit like a yes/no game, but it is definitely clear that different genes and their different variations can have specific effects on athletic performances.

Origins Play a Part in Athletic Performance

The extent to which different studies show a connection between the ACE gene and the ACTN3 gene and an athlete's performance could have something to do with the fact that the comparison included different athletic disciplines. Such heterogeneity increases complexity. It also indicates that athletic performances cannot be explained by one gene, but rather are the result of a combination of different factors.

This diversity of factors is clear when you compare different population groups. It is remarkable that populations perform differently in various athletic disciplines. Asians dominate gymnastics, Eurasian whites do well in power-related disciplines like shot-put, hammer throw and weightlifting, athletes of West Africa are ever present in sprinting and everything from the 800 meters to marathon is dominated by east Africans. These findings have led scientists to study if there are any contributing genetic factors. The fact that Kenyan and Ethiopian athletes performed so well in the long distance disciplines, for example, led to the hypothesis that these athletes have a genetic predisposition that results in better performances than athletes who come from other places.

There is not a single study that could find a genetic explanation for this. Scientists researched whether the 577XX variation of the ACTN3 gene is more present in Kenyan or Ethiopian athletes, but found no clear indications. At the same time they did not find evidence to support the hypothesis that the ACE gene provides an important explanation for their exceptionally great performances. Possibly we need to look for explanations other than genetic ones. Many east Africans are exposed to running at a very young age.

This region possesses many world and Olympic champions, who are role models. The accomplishments of Kim Clijsters and Justine Henin got many youngsters in Belgium to start playing tennis. In the Netherlands the achievements of ice skating champions made the sport popular amongst the youth. The same happened in east Africa with running. Besides, many east Africans live and train at high altitudes, which has a positive effect on running performances. And, last but not least, for some sports it also means hope for a career and economic welfare. Genetic factors do not provide us with a complete explanation for the success of east Africans in long distance running. It seems improbable that there are unique genotypes that explain their success.

Genetic Testing for Athletic Predisposition?

Because professional athletes can make a lot of money, it seems tempting to test young athletes for the known genes. A number of companies do offer genetic tests for the ACE and the ACTN3 gene. It appears to be interesting for some people to determine at a young age, through genetic testing, if power, speed or other athletic aspects are present. This way kids could pick a sport that is best suited for them, in which they would get better results and would give them more satisfaction.

But it is still too early to genetically scout for extraordinary athletic talent. In light of the contradictory research results, the odds on making false conclusions from test result are too great. Besides, identifying athletic talent is a complex process that should not be limited to only the genetic aspect. Technique, tactical and strategic insight and psychological resilience: these are only a handful of aspects that play an important part in being a professional athlete.

Parents know their children best, and that makes them well placed to discover their talents. However, some parents might project their own ideals on their offspring, with all the unpleasant consequences that come with this. A genetic test that determines whether kids are more suited to disciplines that require power or endurance, would give parents an objective reason to push their child into a certain direction. But such a determining test does not exist. If children get to pick a discipline they like to do, it is guaranteed that it will come with better results and will give them more satisfaction than the sport for which 'they are made'.

18
The Godfather

Don Vito Corleone (played by Marlon Brando in the movie The Godfather) passed the leadership of the New York Mafia onto his son Michael (played by Al Pacino) because he seemed destined for it. Is there something like genetic predisposition for criminal behavior?

Different studies that compared criminal records between parents and children found that children of convicted felons often become felons themselves. Can this be explained by genetic factors? Environmental and social factors actually seem to be more influential. Children who are confronted with violent and criminal behavior while growing up will more often consider this behavior as 'normal'. Also, children who grow up in weaker social economic circumstances seem to resort to criminal behavior faster than children who do not know poverty and neglect.

Aggressive and Antisocial Behavior

There is no genetic research that maps out if someone is destined to show criminal behavior. There has been research into genetic factors that might be related to aggressive and antisocial behavior, however. Research from 1993 conducted on a family in which aggressive and antisocial behavior manifested itself and led to criminal behavior (like rape and arson), showed that the males in the family had abnormally low quantities of the protein monoanimin oxidase A (MAOA). This protein plays an important role in regulating the metabolizing process of serotonin, a signal molecule in the brain that influences the communication between brain cells and has an impact on emotions, confidence and sleeping patterns.

Further research conducted in 2002 confirmed the relation between MAOA and antisocial behavior, and showed that children had a higher risk of showing antisocial behavior if they had the variation of the gene that led to lower MAOA quantities. But also if they were neglected or abused. It shows that the genetic component cannot cast predictions, but rather it is the interaction with surrounding factors that is defining when developing a certain behavior.

> **"The genetic component cannot cast predictions; the interaction with surrounding factors is important in developing a certain behavior.**

The so-called 'MAOA gene' is not the only gene that is related to antisocial behavior. On the list of genes related to this kind of behavior, there are also genes that are linked to the production of dopamine, another signal molecule (for example DAT1, DRD2, DRD4) and of serotinin (5HTTLPR), or with the breakdown process of certain molecules (COMT). It is clear that these genes and their variations they might express have an immediate impact on behavior. Genes impact on the production of a number of proteins that influence brain functions and hormones. They are mainly a component in developing antisocial behavior, but never the determining factor.

Predicting Behavior is Not Possible

In contrast to what some headlines want us to believe, a single gene does not determine aggressive or antisocial behavior. Even if there is a genetic connection, it is clear that the behavior is not controlled by one gene, but rather by a mix of different genes, each having their own small impact. Frequently these connections appear on the level of populations; in other words, these results are valid when looking at a large group of people and they cannot simply be converted to predictions about an individual's behavior. Besides, statistical significance does not mean that the connection will actually be socially or clinically relevant. Where science is today, it is impossible to predict if an individual is going to be antisocial or aggressive simply based on their genetic makeup.

The question was raised if these risk factors could be calculated into the procedure to declare an individual insane or partially insane for his or her actions. One well-discussed case is where an Italian judge reduced a sentence because the convict possessed a number of

genetic variations (in his MAOA, COMT, SCL6A4 and DRD4 genes) that are associated with aggressive behavior. The convict suffered from schizophrenia, a mental disorder that leads to hallucinations, chaotic behavior, language use that does not make sense and loss of sense of reality. At the time of the murder the convict was psychotic. Against his doctor's orders the patient stopped taking his anti-psychosis medicine.

Accountability

In order to convict a suspect, the prosecutor has to prove that the suspect is responsible for his or her actions. In criminal courts this often comes down to a discussion about whether or not the suspect can be held accountable for the crime. Psychiatrists are consulted to judge whether or not a suspect is sane. If the culprit is found to be insane, the court can have him forcibly committed to a psychiatric ward rather than sentence him to a jail sentence. In the Italian case, the convict saw his sentence reduced to nine years based on his psychotic schizophrenia. The 1-year sentence reduction on top of that, based on his higher genetic risk of aggressive behavior raises questions. Firstly, science today does not allow us to interpret if an individual who carries specific genetic variations associated with aggressive and antisocial behavior will actually demonstrate this behavior. It is not known to what extent it can be used as a predictive tool. It is problematic to use tests that are still in the research stage in court cases pretending they are tests that have been well validated and that can cast predictions on one's behavior.

Secondly, using genetic factors as mitigating parameters leads to a narrow view on biology. No one wants to get caught up in a deterministic model in which genes are responsible for our actions. It is crucial not to separate one's genetic makeup from their environmental and social factors and to not fall for simplistic causal argumentation.

“Genes do not dictate behavior.

Social Context Remains Important

In the relatively peaceful Western world, criminal behavior is unwanted. Criminals are 'aggressive', 'merciless', 'selfish'. But in times of war this could drastically change to 'persistent', 'efficient' and 'focused on survival'. They describe exactly the same behavior, but in times of war it is no longer unwanted. The characteristics that make people criminals in times of peace could just as well render them war heroes. This is probably the reason why aggressive and antisocial behavior was not weeded out by the process of natural selection over the course of history: in certain circumstances it might come in handy for survival.

Next to aggressive behavior there are other 'less wanted' traits that have a hereditary aspect. Even though today ADHD (*Attention Deficit Hyperactivity Disorder*) is controversial, the behavior it brings with it is associated with a clear hereditary component. It is not at all limited to one gene, but research shows that the DRD4 gene, and more specifically its 7R variation, can influence the form of ADHD in certain cases. According to researchers the variation originated 45,000 years ago, when people were still nomad hunter-gatherers. These days the DRD4-7R genes are more common in population groups that share a longer history of migrations and often still live as nomads. This 'ADHD gene' was clearly not tossed away throughout history, so it does not give its carriers any evolutionary disadvantage.

A fine example comes from Kenya's Ariaal tribe, a group of shepherds that herds camels, cows, sheep and goats. The tribe lives in the desert and is constantly on the move, searching for food for its livestock. Young males that carry the DRD4-7R gene mutation are the better shepherds of their tribe: they are more alert than the males without this variation, and are tireless. They also constantly scan their surroundings for signs of predators, changes in weather, or a lack of food and water. Today there are members of the tribe that gave up their nomadic existence and are committed to agriculture, sell their harvest on local markets and send their children to school. Research shows that the men from the Ariaal tribe carrying the DRD4-7R variation that remained living the nomad lifestyle are less underweight than their tribe members without the variation. In the case of the sedentary men of the Ariaal it is the other way around: males with the DRD4-7R gene are more underweight than their tribe members that do not carry the variation.

This does not form a hard conclusion, but the research could show that people with the DRD4-7R gene are more successful in a nomadic existence. Imagine that we transformed this into our society: if you put these boys that have a genetic makeup for herding cattle on a school bench for an entire day, the chances are they will not be able to keep quiet. In any case, the study performed with the Ariaal tribe is a clear indication that it is mainly the social context that determines if ADHD is a 'blessing' or a 'curse'.

On the one hand it is true that some individuals showing aggressive or antisocial behavior possess some of these genetic variations. On the other hand it is also important to realize that a lot of other individuals possess the same variations yet do not show any signs of these behaviors.

Labeling people with an 'aggression gene' is scientifically unfounded and only leads to stigmatization. Besides, saying that a gene is what is causing this behavior will give some an excuse to act that way. Some might interpret this solely as a responsibility issue. Others might find solace in punishing these people harder because there is a higher chance of relapse. This unjust use of genetic information can impact on an important legal principle, like the fact that one is innocent until proven guilty. Instead of being innocent until proven guilty, possession of certain genetic variations can lead to a suspicion of guilt.

There are no genetic variations that could have predicted Michael Corleone taking over his father's business as the head of the New York mob. At the most, Corleone Jr. possesses some of the genetic variations associated with antisocial and aggressive behavior. And if his father carried them as well, he has a higher risk of inheriting some of those. But genes do not dictate behavior, or at least not in a way that made the Corleones excel at what they did.

19
Personalizing Medicine to Fit Your Genetic Profile

Your body's reaction to certain medicines is determined by your genetic profile. Would it be useful to perform genetic tests before prescribing medicine?

Some medicines will work better or worse depending on genetic factors. Some people will break down some medications too fast, which does not give them enough time to properly do their job. In other cases, they will form a toxic by-product that is harmful to the patient and could even cause death. Just for that, it could be useful to perform a specific genetic test before prescribing certain medications.

But it goes even further than that. New, tailored drugs will only work on patients with a certain form of cancer and not on others. The same goes for patients who suffer from a specific form of hereditary dementia, but not other types of dementia, even though on a clinical level you would not be able to tell them apart. The pharmaceutical industry is trying hard to develop these medicines, but what they are actually doing is tracking down the right patient group.

Allergies

Do you know anyone who has an allergic reaction to some medication? Or did you ever develop a strange reaction to a medicine yourself? Maybe sometimes you think: this is just not working?

The answer to all those questions is that people can have different reactions to drugs, and this is for various reasons. Of course you have the typical circumstances to respect, which everyone is familiar with and that the pharmacist will point out to you: to take before or after dinner, respect the dose, do not combine with other medication or alcohol. Sometimes the dose may vary according to age and weight, but also with reactions in the liver and/or kidneys, for example. Aside from that, there are plenty of other differences that can cause medicine to work for some and fail for others. Genetic factors play a role in this.

Pharmacogenetics and Pharmacogenomics: Overlapping Disciplines

Pharmacogenetics studies the different reactions individuals might have to certain medicine: researchers try to make a connection between the reaction of an individual to certain medication and his/her genetic constitution. The difference can be related to the speed of converting medication to its active component, but also to how fast that active component is broken down and thus becomes inactive.

> **“That is why you want to offer a patient a pharmacogenetic test before giving medication and not vice versa, especially when you are expecting side effects.**

Both processes not only influence the therapeutic effect of the medicine, but also change the side effects. Researchers divide large groups of patients based on the genetic variation they discover rather than try to predict the response of an individual patient to a certain medicine based on the finding of such studies. That is why you want to offer the patient a pharmacogenetic test before giving medication and not vice versa, especially when you are expecting side effects. This all happened as scientists became convinced of the influence that genetic variations have on the medicine's interaction with the body. Because of new genomic techniques and because genetic information is more easily accessible nowadays, a new scientific discipline emerged: pharmacogenomics. Those two disciplines overlap and essentially can be broken down to the same thing: mapping genetic differences that can influence the effects or side effects of medication, and, if necessary, suggest the patient have a genetic test before prescribing medication.

Let's Get Personal

This field is sometimes known as personalized medicine, which sounds like a buzzword, if anything. But it does hold some truth: if medical advances keep progressing and pharmacogenomics delivers the expected results, doctors will be able to prescribe treatment based on the genetic constitution of their patients.

Is this the soundtrack of the future, you may ask. Not quite. Already there are important – and rather easy to perform – genetic tests available that can save lives. Besides, there is medication available (and there are even more in the pipeline) for clinical studies of which the effect is almost entirely connected to a genetic variant. These are known as *drugs designed for you* or *prescription medication tailored to an individual's genotype*. Such drugs will not be given lightly. Those medications will only be used on people who stand to gain from it and where serious side effects will be avoided.

The Various Aspects of Genes and Medication

When you talk about genetic tests and corresponding medication, you have to differentiate between two actions: 1) managing the dose in order to minimize side effects and maximize the effect of the active component; and 2) looking for the right patient to take the right medication and vice versa. In practice, the approach will be fairly similar because the two aspects are related and the lab techniques that are used for genetic testing are in a way comparable.

Pharmacogenetics studies the differences genetic variations might have on side effects and the breakdown period of specific medications. Not all medication is susceptible to major individual differences when it comes to the speed it's broken down and its effectiveness: it depends on their chemical class and the path through which the body processes and breaks down the medicine. There are a couple of groups of medication that do present major individual differences in patients, and it has been proven that this has to do with the genetic build-up of the corresponding enzymes. You can compare this with the Asperger's example we gave in Chapter 4.

From a clinical perspective, we usually refer to the drug Imuran because the way the medication breaks down in different people can lead to serious side effects. The medication dosage needs to be adjusted to each patient based on the presence or absence of certain genes. Another example is the connection between the specific HLAB*5701 variant in the immune system and oversensitivity to Abacavir, a medicine used in different combinations to treat AIDS. The genetic test enables us to estimate the chance of having side effects. The variant's frequency varies according to different groups, but the side effects are the same: fever, nausea, vomiting, diarrhea, and some patients get rashes on their skin.

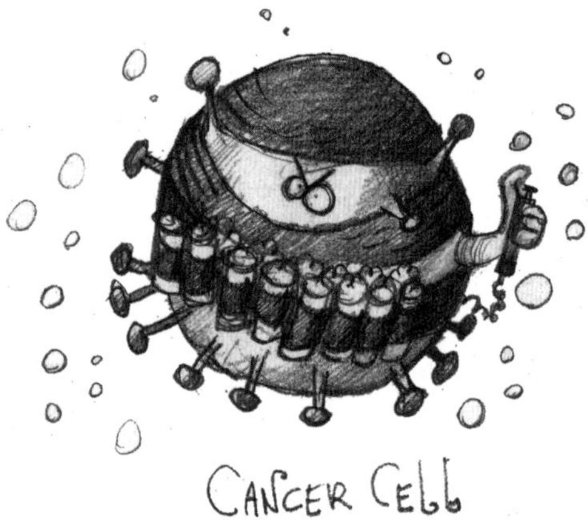

Other tests that have been claimed to be clinically important and that sometimes get promoted are tests that analyze the different cytochrome or CYP genes (like P450). Those genes play a role in some medications including a number of neuroleptics, which are used to treat psychosis and syndromes linked to hallucinations and psychomotoric agitation. Even though it sounds promising to be able to determine the right dose based on genetic testing, there are still a lot of doctors who are more inclined to experiment with the dosage patients take in order to reach the desired effect, or just prescribe a different medication. This behavior can partially be justified since the test results are not always conclusive. At the same time, the effect of getting the dosage right from the start is heavily underestimated: if a patient is prescribed the right dosage immediately, there is a larger chance that the patient will feel the difference and will continue the treatment.

Pharmacogenomics includes the study of diagnostic and prognostic variants that could be used in clinical practice. Some genetic variants, for example, are helpful in categorizing tumors, and by consequence in targeting specific drugs. Also in the field of dementia genetic tests are increasingly being used to determine the type of dementia that a patient is suffering from, which makes it possible to estimate the disease's course. It looks like the choice of medication will soon be linked to this genetic information.

The best-known example is the use of Herceptin for breast cancer. This medication is actually an antibody used against HER-2-proteins that can be found on the cellular walls of some cancer cells. This antibody binds the HER-2 and prevents this protein from getting signals that it can pass to the cancer cell. The medication only works when the cancer cell actually possesses HER-2, which is the case in only 15% to 20% of breast tumors. Genetic testing enables us to see whether or not the cancer, or more accurately the patient, is eligible for treatment with Herceptin. This choice is important on the one hand because HER-2 positive cancers are aggressive, thus requiring adequate treatment, and, on the other hand, because the treatment is very expensive so there is no need to use it if you know that it is not going to work beforehand. Some countries have linked reimbursement to the result of the genetic test: only if the genetic test is positive will the medication be reimbursed.

> **“Genetic tests are useful to confirm the diagnosis and start treatment.**

Most other examples from clinical practice come from the domain of cancer treatment. Genetic tests will track genes like KRAS, ALK1 and EGFR for genetic variants in view of the treatment of certain types of lung cancer. To treat chronic myeloid leukemia (CML) doctors can prescribe Imatinib (Gleevec) when this blood cancer is caused by Philadelphia chromosome, a genetic defect where Chromosomes 9 and 22 stick together and cause an oncogene (a gene with the potential to cause cancer) in that specific place to turn on (see Chapter 5). The genetic test is useful to confirm this diagnosis and start the therapy. There are even more combinations of genes and medication in the pipeline, including for MEK, BRAF and PIK3CA.

The Principle is Not Entirely New

To be clear, the use of diagnostic and prognostic testing is not new: in pathology researchers have for a while been using histological and biochemical markers to determine the malicious nature of a tumor,

TMPT and Imuran, a Useful Combination

Imuran is a medicine that is used to treat autoimmune diseases, such as Crohn's disease and specific cases of rheumatoid arthritis, and is oftentimes also prescribed after organ transplants. The possible toxicity of the active component, Azathioprine, is partially determined by the by-products that are formed when the body is processing this medicine. Thiopurine Methyltransferase (TPMT) is one of the enzymes Azathioprine meets as it is processed. Variants in the gene that cause the coding for TPMT result in the underperformance of the enzyme. In some cases, it does not even perform at all. In this case, the Azathioprine is mainly processed into by-products that suppress the activity of the bone marrow (the making of white blood cells) or damage the liver. In 0.3% of the population, the activity of this enzyme is pretty much non-existent, and in 11% of the population the enzyme only partially works. Thanks to genetic testing, doctors can determine if a patient will have a higher chance of having side effects. We do not yet know all the possible side effects, and allergic reactions and pancreatitis in therapy with Azathioprine are not always linked to the presence of these known TPMT varieties. There are other factors that play a role which causes this.

which allows them to fine-tune a clinical approach and treatment accordingly. This scientific field is expanding exponentially with genetic and genomic analysis.

Similar studies investigate the treatment of other diseases, including Alzheimer's. Sometimes the approach works, but not always. The starting point, however, stays the same: if the genetic cause of a patient's condition is better understood and a treatment can be developed that corresponds to the defects in that specific gene, the chances of success with that patient are greater. The correct term is *companion testing*: the genetic test accompanies that medication or – to be completely correct – medical doctors start with a genetic test and depending on the result the medication is or is not prescribed.

“A total genome analysis of tumor material cannot be taken lightly

The American Food and Drug Administration (FDA) is an agency that controls the commercialization of medication in the United States of America and is quite strict when it comes to pharmacogenetic testing of certain medication. The FDA decides when pharmacogenetic testing is mandatory or recommended, and this is for safety and performance reasons. Internationally most countries are more or less in agreement when it comes to these procedures. But since reimbursement of the medication and the genetic test is not the same for all countries, implementation might differ from county to country.

Should the Entire Genome of the Cancer Be Mapped out from the Start?

Pharmacogenetics and pharmacogenomics are both fast-growing fields. The combinations between the better understanding of the cancer's origin and the estimation of the patient's reaction to the treatment are interesting. It is clear that a specific mutation of the cancer genes in the patient's tumor can guide the choice of medication, but also that the rest of the patient's genome can contain genetic variations that can influence the chance of a positive response to the treatment. Since a genome analysis of the tumor immediately gives the entire genetic constitution of the patient, cancer research is really being pushed in this direction. Technically speaking it is possible, however, the reimbursement of this procedure has not yet been fine-tuned, but this will probably be fixed.

It is important to realize that when a genome of a tumor is sequenced, the entire patient's genome is known. It is about his or her cells. You could almost state that this knowledge is a by-product of the analysis. All aspects of a genome analysis are applicable to this case: the patient can get information about the other genetic variations, called incidental findings (we talked about this in Chapter 12). This brings in a whole set of ethical issues.

Microbiome

The microbiome is the sum of all micro-organisms (bacteria) which live in and on our bodies and accompany us constantly. Most of these organisms live in good cooperation with the host: think about the bacteria in our digestive system that optimize digestion. Our skin houses millions, even billions of bacteria and the composition of these colonies differs depending on the place on our body: underarms, pubic area etc.

The importance of the micro-organisms who live in complete harmony with humans has been greatly underestimated to this day. New techniques of DNA sequencing enable the complete mapping of a human's entire microbiome. This has not yet been applied in clinical practice but it is being researched. Big companies in the food industry have developed a huge interest in this: what you put in your yoghurt can greatly affect a human's microbiome.

The expectations are that many connections can be discovered between illness and health on the one hand, and the collections of micro-organisms on the other hand. It mainly concerns the illnesses such as auto immune deficiencies such as diabetes, rheumatoid arthritis, MS and maybe even obesity. Some predict that these kinds of environmental effects can even influence neurological problems, such as depression and schizophrenia, but it is too early to make conclusions about this. The question is when microbiome sequencing will be introduced in hospitals and lifestyle centers.

A By-Product of Genome Analysis

Genome analysis in clinical studies is important to detect the genetic variations that play a predominant role in pharmacogenetics and pharmacogenomics. But it would be too premature to perform complete genome analysis on all the patients as soon as they go to hospital to prevent any unwanted side effects of medication.

Specific Treatment When It Comes to Cancer

Breast cancer is a field in which the patient's DNA is becoming more important every day for finding or fine-tuning treatment. This happens in two ways.

On the one hand researchers look for genetic variations on the tumor itself which give away clues to its origin. Depending on this result the treatment can be modified. The best known example is that of the HER-2 protein and its inhibitor Herceptin.

On the other hand the genetic constitution of the patient (and thus not only the tumor) can give hints on how to successfully treat it. An example is the PARP1-inhibitors that are currently in clinical trials at the moment. PARP1 is a protein that plays a role in the recovery of DNA damage, just like BRCA1 and BRCA2 (see Chapter 5). Research has shown that cancer cells with a BRCA1 or BRCA2 defect use the PARP1-path to escape the toxic effects of cancer medication. That is why the pharmaceutical industry has been successfully testing medication that suppresses PARP1. There is biological and clinical evidence that this double strategy works: patients with a genetic defect in BRCA1 and BRCA2 react better to a combination of chemotherapy and PARP1-inhibitors than to the classical treatment. It comes down to figuring out whether or not the cancer has a BRCA1 or BRCA2 defect in order to optimize the treatment. More specifically this means that soon every woman with ovarian cancer will be tested for mutations in the BRCA1 or BRCA2 gene, and this will not just be limited to women who are eligible because of their family history. Eventually this procedure will also be used to treat breast cancer, but the research is still ongoing.

If this field keeps expanding soon the genome of every cancer –or at least a part of it– will be mapped out. And because sometimes cancer changes its strategy like a chameleon changes color, it looks as though individual cancer patients will require new testing in the different stages of the treatment. It will become a big market with a lot of commercial importance. The correct systems for making sure patients will be reimbursed for all that genetic testing have to be set up urgently.

It is important however to start setting up systems that make it possible – in case a patient undergoes a genome analysis for a specific clinical problem – to collect all the useful and relevant pharmacogenetic data from that genome. This is not science fiction. In the near future these 'apps' will be active in every medical genomics data control system. An argument to rapidly release genome analysis when it comes to the research of tumor material is the limited availability of the material: from a small biopsy of the tumor only a little bit of DNA is isolated. If the same material allows you to either determine one or millions of genetic variations, preference goes to this second option. Of course the test has to stay affordable, and ways should be found to filter it appropriately.

20
Gene Therapy, a Medicine?

Gene therapy was the great promise of the medical world for the last quarter of the last century. At this time its success is limited, but there is hope for the future.

In a limited number of cases, repairing genes can be done clinically. The limitations are technical (how do you fix a piece of DNA in a living cell, and in all living cells of the body?), ethical (should we 'modify' egg cells, sperm cells or embryos?), and also clinical (not all problems can be fixed by replacing the gene, i.e. in many cases the damage has already been done).

Gene Therapy, Thirty Years in the Pipeline

Correcting genes, replacing genes, turning off genes. It makes some people think about science fiction, whereas others ask why it is not possible: just replace the defective gene and the problem is solved. Reality is somewhere in between: it is not easy, but it is no longer science fiction. The development of gene therapy is faced with many technical, biological and medical problems, which after more than 30 years of research have led to only some treatments reaching the point of clinical application.

The principle is simple and seems applicable to any genetic defect. There was hope that this form of therapy could be used easily to treat hereditary disorders by inserting a healthy gene into a cell to replace the defective gene. But the obstacles are great and hard to overcome. The intact gene needs to be placed in a vector – a type of shuttle – and needs to end up in the cell one way or another.

> **“There have been fantastic experiments that have been executed with great care and have never led to accidents.**

Some viruses are extremely skilled in infecting cells and integrating their viral DNA into the cells' genome. That is why scientists modified 'wild' viruses that could keep their ability to penetrate a cell, carry with them the required 'cargo' and deposit it in the genome of the host cell. Scientists tested many different viruses and the HIV virus was one of them. There have been fantastic experiments that have been executed with great care and have never led to accidents.

Prudent Success, but Not in All Cases

The next obstacle is trickier: the gene needs to end up in the right cells, in the right organs, such as a nerve cell in the brain to treat a genetic form of Parkinson's disease, or inside the inner walls of the pulmonary alveoli to treat cystic fibrosis. This last case made it clear how problematic the technique really is: the transfer of the healthy gene was a success: the DNA was deposited in the cell's nucleus but it did not integrate into the genome of the patient's cells. The result was that the healthy DNA was useable for only a short amount of time (a couple of days or a couple of weeks) before the repaired cell disappeared. This is not a problem if you can repeat the process every other week. But because the patient's immune system treats the vector as if it is an unknown assailant – like any other virus – there was soon resistance, not against the correct gene, but against its 'shuttle'. This shows the biggest problem: if the preferred correct gene is built into the genome of a patient's cells, it should be in a place where it cannot cause extra damage. Some adverse events happened and they show how delicate gene therapy is.

> **“It is about a small number of patients, making gene therapy extra expensive.**

Finally, gene therapy for somatic purposes is only to be considered when you want to treat patients who are still able to heal. The damage done by the disorder needs to be reversible and you need to be able to intervene early in the disease's process. This is not a given for many congenital disorders.

Examples of Success Stories

A case that made the news is the treatment by gene therapy of Leber's congenital amaurosis (LCA), a hereditary eye disorder. LCA is rare: approximately one in 80,000 children are born with this disorder. In most cases these children have problems with their eyesight at birth or shortly after, which leads to blindness, and also comes with some other, sometimes serious, brain disorders. There have been many different forms of this disease that have been described – almost 20 – and at the moment they have found 15 genes that are at the base of one or many of these forms. A small number of patients are involved, making gene therapy extra expensive. The whole procedure, going from clinical protocol, safety studies to clinical studies, needs to be set up for a rather small number of patients needing treatment. The success story to this day has been limited to LCA2m, a recessive disorder that is caused by defects in the RPE65 gene. Through a modified virus with the correct gene, which is put directly into the eye, some patients' eyesight has been restored without any indication of side effects. In the meantime, this has demonstrated that the earlier the procedure is undertaken the higher the rate of success. This story brings hope for the patients and families affected by LCA2, but also to others who are going to have to wait until it becomes clear if other forms of this disorder can be treated with the same success.

Cystic Fibrosis: Gene Therapy Has Not Been a Success

A lot has been invested in finding a workable gene therapy for cystic fibrosis, the most frequent recessive genetic disorder. The first studies were published in 1990.

A patient with the disorder will usually have abnormal functioning of the CFTR protein. CFTR is the name of the cystic fibrosis gene and the corresponding protein. This protein can usually be found inside the cell's wall and forms a canal that the cell uses to pump out chloride ions. If it were possible to integrate the correct gene into the cell, this would enable the cell to produce normal CFTR proteins, send them to the membrane and give itself a chloride pump. With cystic fibrosis, the lungs and pancreas are the main targets, and it is usually problems in the lungs that lead to early death. Unfortunately gene therapy has yet to succeed in allowing the lungs to take in the healthy gene with high-efficiency. One of the reasons is that the lungs fire off a defense mechanism against the viral 'vector' and its cargo, making gene therapy practically impossible. Researchers are now trying other methods to introduce the CFTR gene in to the cells rather than via the viral molecules. One way is to use microscopic balls of fat that hold the DNA of CFTR, and that can be absorbed by cells. The efficiency appears to be too low to use this method in a clinical context. Scientists have not given up hope but it could take years before gene therapy can be applied successfully to cystic fibrosis.

In the meantime, researchers are making progress on the treatment of cystic fibrosis through medicine that alters the process or can reactivate the poorly working chloride pump at the molecular level. There is hope for these patients, despite all the current disappointment of gene therapy.

In 2012, American and European authorities approved a gene-therapy-based treatment for a rare form of obesity. Patients with a defect in the lipoprotein lipase gene (LPL) show severe, sometimes life-threatening attacks of pancreatitis (inflammation of the pancreas), even when they stick to a strict low-fat diet. They can now be treated with Glybera, which is made from a solution containing a modified virus that carries the normal LPL gene (they named the new virus molecule *alipogene tiparvocec*). After the patient is injected with the virus, their muscle tissue starts producing its own LPL protein. Next it goes into the patients' bloodstream and eventually their pancreas where it helps to break down fat. Of course this is not a medicine that you can just get at the pharmacy. It is reserved for special cases, namely patients who have the hereditary form of PLP deficiency and do not respond to any other form of treatment.

There are also descriptions of gene therapy treatment for other disorders, including some hematologic disorders (chronic blood cancer and acute lymphatic leukemia, CLL and ALL respectively), for hemophilia (a blood-coagulation disorder), one or two congenital immune system disorders and even for Parkinson's disease. Everything depends on the kind of defect and the tissue in which the correct gene needs to perform. For blood cells this is relatively feasible, as it is also for the retina, liver and specific areas in the brain too. For other organs, like the heart, there are still a lot of questions.

Defeats and New Challenges

The road to gene therapy started in the 1970s, but it wasn't until 1990 that the first clinical study was approved in the United States. It was for the treatment of the so-called 'bubble boys', children with SCID (severe combined immunodeficiency), a congenital immune system defect.

Shortly after that, other clinical studies followed for a number of congenital metabolism disorders. It started out successfully, but in 1999 18-year-old Jesse Gelsinger died during treatment. He suffered from a rare disorder that, if untreated, could lead to a dangerous build-up of ammonia in the body. Gelsinger could control his disease with medicine and a low protein diet, but he really wanted to be cured by gene therapy. He received a very high dose and his immune system reacted destructively towards the weakened virus that was used in the study. He died four days after injection.

“Gelsinger could control his disease with medicine and a low protein diet, but he really wanted to be cured by gene therapy.

In France, scientists conducted a lot of research into the treatment of congenital immune system defects, including SCID. Also in 1999, 11 patients were being treated. They each got their own bone marrow transplants, which had been treated in a lab so that the gene in these cells had been corrected.

For nine patients the treatment was successful; two other patients were diagnosed with leukemia. This blood cancer was caused by the fact that the virus containing the correct gene nested in the genome of certain cells, in the middle of an oncogene. Because of this setback, a number of clinical studies were suspended and were only partially resurrected years later.

Do Not Give up Hope!

Fortunately, over the years there have been a number of treatments that actually reached the clinical trial stage in hospitals. The field continues to expand and there are still different clinical studies ongoing for hereditary disorders and cancer.

Very recently researchers have discovered a mechanism in bacteria that enables the copying and pasting of pieces of DNA directly into the cell. This is called the CRISPR/Cas9 gene editing mechanism, and it is now used by many researchers in laboratories to modify cell DNA and to turn genes on and off. If all this research shows that the mechanisms work flawlessly and can be applied to the patient's cells or directly into the patient, there is new hope for success. But it does take years before such discoveries can be converted into clinical applications.

Restoring Genes or People?

Gene therapy up to this point had been directed at specific tissues and organs. Why is it not possible to just restore the entire human being in one go? In theory this is possible if the correct gene is restored in the fertilized egg cell before a new embryo, and thus a new human, starts to develop. In this case we are dealing with germline transfers or germline corrections. In most countries there is a moratorium on germline transfers: the techniques are not yet safe enough for the parents' egg cells or sperm cells to be modified without causing a defect.

Correcting the Defect, Not the Gene

There are also different ways to correct a genetic defect. Researchers can also try to intervene in the defect caused by the gene. Strictly speaking this is not gene therapy because it does not require inserting a healthy gene; instead, the incorrect gene is stimulated to work properly. Practically speaking, it comes down to this: researchers try to stick tiny molecules (based on DNA, but with a different chemical composition to be more stable) to the incorrect gene. The idea is that the cell's machinery will ignore the mistake and will 'overwrite' the gene in order to make it into a protein. This works well with cultured cells in laboratories, but is less efficient in the patient's living cells.

> **“The normal enzyme missing in the patient can be produced in the laboratory and can be administered to the patient.**

Many companies have been involved in the development of these 'Band-Aids' for genes. An example is therapy with so-called *antisense oligonucleotides* in patients who suffer from Duchenne disease, a severe muscular disorder. These molecules make sure a piece of the genetic code within the mutated Duchenne gene is ignored so that an almost-normal protein can be produced in the muscle fibers. These molecules were first tested by a local injection in the patient's lower leg muscle. The results were promising: the proteins that once were missing in the patient's muscles were now detectable at the injection site. Later they came up with a general treatment through subcutaneous injections, with very encouraging results in different smaller studies. A greater clinical study, however, could not confirm the expected slowdown of the disease's evolution. After a year it became clear that the distances treated and untreated patients could walk were the same. Further research seems to show that earlier and longer treatments could lead to better results. This principle has since been used for different disorders, and there is hope that something similar might be possible for cancer treatment.

Dolly, A Crash Course in Cloning

A blue tomato, a fluorescent begonia: it is possible but the 'costs' are high. Researchers made many deviant tomatoes and ugly begonias before they could call the experiment a success. This is also the case with cloning animals: the famous example is Dolly, the cloned sheep from 1996. It was great news when the scientists at the Roslin Institute in Edinburgh announced in 1997 that they had cloned a sheep. Cloning happens by taking the nucleus of an adult cell (from an adult body) and putting it inside a fertilized egg cell, from which the nucleus has been removed. By using a specific treatment scientists could reprogram the adult cell's nucleus and develop as if it were the cell nucleus of a recently fertilized egg cell. This is necessary, because all of the genetic information needs to be read in the correct order for the normal development of the embryo (see Chapter 3). The newly composed egg cell is then implanted into a mother's womb, to further develop until the time of birth. This was successful in 1996 using a sheep. What is often left out is that Dolly was the only sheep that was relatively healthy and was delivered to this world appearing to be intact. In the experiment that eventually gave birth to Dolly, originally nearly 300 egg cells were treated. Less than 30 of them grew in the petri dish until they could be placed back into 13 surrogate mothers. Seven sheep were impregnated and two pregnancies failed leaving 5 lambs to be born. Two immediately died after birth, and a third one ten days later. Why Dolly was found to be the newsworthy one of the two remaining clones cannot be clarified by the scientific publications. But the conclusion is that most cloned animals do not make it to this world or cannot live healthy lives.

DOLLY, THE CLO(W)NED SHEEP

This is, next to the ethical and social questions that arise when discussing cloning humans, the most important technical reason to forget about this and not even start. Cloning will be successful in the animal world: copying a perfect race horse can make a lot of money, even though at this time nothing can guarantee that the clone will race just as well as the original (and we won't know, as cloned horses are banned from racing...). Besides, we are already familiar with clones and know that they are not necessarily perfectly identical: they are identical twins. After Dolly, scientists have cloned other sheep, including four identical to Dolly. In the meantime they have also started cloning rats, goats, cows, pigs, cats, dogs, monkeys and mice, even though that last one came with a scandal: the first publication about it appeared to be falsified. Up to now scientists have not yet cloned primates or humans.

Dolly died in 2003 due to a lung infection, combined with premature arthritis. It is not sure if the disease can be tied to the cloning. One of the main questions about cloning is whether or not the clone receives a piece of the original individual's age or if they are actually starting from scratch when the clone starts its life cycle. Indeed, our cells age as we age – or vice versa, we age because our cells age and the biochemical processes and structures in our bodies become run down.

The reason behind the name 'Dolly'? Researchers used a cell from the mammary tissue from an adult ewe to harvest the cell's nucleus from which the sheep was made. The researchers in Edinburgh therefore decided to name the sheep after country singer Dolly Parton. Any further resemblance is purely coincidental.

Cloning with the deliberate goal to produce genetically identical individuals is called reproductive cloning. In contrast, therapeutic cloning has not the goal to produce an entire organism, but rather cells and tissues of a patient. This form of cloning, which involves stem cells, has an important future in the medical world but will not be discussed further in this book. Stem cells have a connection with genetics, but this topic is just a bit too far removed from the theme of this book, namely genes and genomes.

Let's Administer the Protein

Originally the most attention was given to the treatment of metabolic disorders caused by the lack of a certain protein, in most cases an enzyme, in the patient due to a hereditary defect. There are a wide variety of such disorders, and they are all extremely rare. The market for this kind of medication is very small. At the same time, success rates in the form of a permanent cure are not equally great for all applications. This causes these treatments to be very expensive. The most important criteria for setting up research and clinical studies are mainly the same as the ones for the other applications: 1) the ability to reach the tissue in need of the enzyme to cure the patient; and 2) the extent to which the admittance of the enzyme will actually mean improvement to the patient.

The approach is called enzyme replacement: the normal enzyme is missing in the patient but can be produced in laboratories and can be administered. Again, this sounds simple, but it is not. For example, enzymes can be extremely unstable and they tend to be large molecules so you cannot just chase them through the cell's wall. The approach is based on knowledge of genetics, cell biology and biochemistry: researchers should not only know what the gene defect is, but they also need to know how the enzyme functions in the body, how to stabilize it and where in the cell it should end up. Also for regular treatment of patients they need to create large quantities of enzymes – even if annually only a couple hundred patients worldwide use them. To give you an idea: this is done in 'incubators' in which the cells that produce the enzyme are cultivated. You can compare them to the fermentation vessels in a modern brewery. After production, the enzyme is purified and finished off and has to line up for a number of quality checks.

For some rare, lysosomal storage disorders (LSDs), like Fabry's disease, Hunter's syndrome, and Gaucher's disease there are currently medicines available on the market. In these cases we are dealing with enzymes that are absorbed pretty easily by the cell. They end up in small compartments (lysosomes) in which they should operate. They play a role in detailing and processing other proteins in exactly those compartments. This is a coincidence that makes this group of disorders fairly easy to treat.

An easier example is the factor 8 protein that is used on hemophilia A patients because of a defect in a gene that produces an abnormal protein, if any. Factor 8 is usually produced by the liver and introduced to the bloodstream where it plays a role in blood coagulation and the healing process of wounds. This means it does not have to face any obstacle when the external factor 8 product is introduced directly in the blood.

Mitochondria and other cell compartments are much more difficult to get to, making these treatments unavailable for disorders in these parts. There are also additional obstacles if the product needs to reach the brain. That is why not all metabolic disorders can be treated by enzyme replacement.

Orphan Diseases, Orphan Drugs

Thanks to efforts of patient organizations for rare diseases and political decisions on the European level, the development of therapies for rare disorders received a special statute. Such therapies are called 'orphan drugs' because rare disorders were often referred to as 'orphan diseases' due to the fact that for a long time they did not receive attention. It is better now, but the limited numbers cause drugs to be very expensive once they eventually make it to the market, very much in contrast to blockbuster medications.

To sum it up, the knowledge of genes and the ability to read an individual patient's genome looking for the cause of his or her hereditary condition have led to big breakthroughs in the diagnostics of these conditions. What has to follow next is the effort to also develop them into therapy for this broad spectrum of specific genetic defects. Thanks to initiatives of the European Commission, things are advancing. But the road to discovery of a new genetic defect is generally speaking a long one, not mentioning some successful exceptions.

21
Consumer Genomics

On the internet you are only a few clicks away from ordering a genetic test. But how useful are these results?

You order and pay for a genetic test on the internet, the company sends you a sample kit and you use the cotton swab from the test to collect some saliva from the inside of your cheek (or you spit into a test tube). You send it back and a couple of weeks later you receive an email with the results or can access them on a secure website. That is how easily you can determine a part of your genetic profile these days. But how reliable are these results? And can you interpret them yourself? Who can you go to if the results scare you, or are incomprehensible?

More and More Genetic Tests Available on the Internet

Genetic tests are usually administered within a health care setting, supervised by medical professionals. But nowadays, commercial companies have started to offer genetic tests, oftentimes without involvement of a doctor. Although there might be no companies offering such services in your country, the internet makes such services available without borders. The offer ranges from pure quacks to scientifically sound services. Here are some examples of what is being offered: a test that finds a love partner that would suit you genetically; a test that indicates how a medicine will react based on your genome (see Chapter 19); a test that calculates your risk of being bald; a paternity test; a test that determines whether

or not you are a carrier of an autosomal recessive disorder (see Chapter 7) or a test that assesses your risk of 100 frequent and less frequent common disorders (such as osteoporosis, prostate cancer or type 2 diabetes) by comparing your risk level to the average level in the general population. You can also get information about your ethnic origins, degree of relatedness, information about the color of your eyes, the consistency of your earwax, etc.

The Consumer's Right to Know His Own Genetic Profile?

The companies' marketing strategies focus on the consumer's right to genetic information: 'your health, your right to know' and 'the only people who should know your genetic information are yourself and the people you want to share that information with'. Most companies focus on the confidentiality of the matter and respect for your privacy: they send the results directly to the consumer. So this information does not end up in the individual's medical file. These companies also claim that knowledge of your genetic profile leads to better health: in case of a negative result you will not be worried, and in case of a positive result you can take preventive measures like changing your lifestyle.

The Fine Print: A Lot Less Promising

"The services are for research, informational, and educational purposes only. The services are not intended to be used by the customer for any diagnostic purpose and are not a substitute for professional medical advice." "The information on the website is for informational and educational purposes and is not intended to be used for medical advice or diagnosis or treatment." Almost all of the companies include a paragraph in their general conditions that waters down the quality of the test and that contradicts what is promised in the advertisements.

Different scientific studies that have examined and evaluated these tests all point to the lack of predictive value of the test results. This is particularly the case for tests that offer 'personalized' nutrition supplements and vitamins to the consumer, usually above market prices.

Should You Form a Conclusion from the Results of These Genetic Tests Yourself? The Risks of Risks...

Ok, so you have received the results of a genetic test. Now what? What does this information have to offer? Can you yourself draw conclusions from it with regards to your health? In most cases these companies will offer risk predictions for multifactorial conditions. Not only genes but also environmental factors and a component of (bad) luck play a part in the development of such disorders.

> **"If you buy the same test from different companies, you will sometimes end up with contradictory results.**

Imagine that the genetic test results show that your chance of contracting type 2 diabetes is 10%. On average 5% of the population will develop type 2 diabetes, thus the 'normal' risk is 5%, and your risk thus seems to be twice as high. The risk is tied to genetic or hereditary factors, but our lifestyle also has an important influence: a life sitting down, bad eating habits and obesity facilitate the development of diabetes. The risk of this disease can be better determined by a doctor asking specific questions that gage for your waist size, BMI, nutrition and exercise habits, etc. than a genetic test can. Watching your weight, exercising a lot and reducing red meat intake but increasing fruit and vegetable intake will be the health advice you get, regardless of the results of a genetic test.

Besides, you are supposed to take this genetically determined risk factor with a grain of salt. If you buy the same test with different companies you will sometimes end up with contradicting results: higher risk with some companies, lower with others. Companies choose different genetic variations, make different calculations and compare your risk against different data. Besides, a statistically higher risk does not make it clinically useful.

From Invention of the Year 2008 (TIME Magazine) to a Prohibition on Offering Genetic Tests in 2013

23andMe (www.23andme.com) was one of the first companies to offer consumers an extensive genetic test over the internet. In 2008 the company received the *TIME magazine*'s Invention of the Year award. The company tested individual consumer's DNA against roughly 1 million genetic variations and a couple of hundred disorders and made a comparison of the risk of the test subject and the average risk of the population.

This became a success story: it did not take very long for 23andMe to become a market leader, while the test price dropped from 750 EUR to 75 EUR in only 6 years. In 2013 a spokesperson for the company declared that the focus was on growth, not profit: 'Growth needs to push us in the direction of profit.'

The capital for this company came from different investors including Google Ventures and pharmaceutical giants Johnson & Johnson and Roche. The company has a strong tie to Google: the founder and CEO of 23andMe, Anne Wojcicki, used to be married to Serget Brin, co-founder of Google. Commentators suspect that companies like 23andMe want to search the genetic data and health information provided by consumers to look for interesting research results that could lead to profitable diagnostic and therapeutic applications or in other words: genetic 'consumer profiles'.

But this nice story temporarily ended in November 2013 because, after years of hesitation, the US Food and Drug Administration (FDA) prohibited 23andMe from selling genetic testing kits for health purposes to individual consumers. Doubts about the clinical value and analytic quality of the test were important arguments. Besides that, the FDA also thought it was irresponsible to give the consumer information about a couple of hundred possible disorders but hardly any advice about what they should do with this information. The lack of medical follow up for the consumer convinced the FDA to prohibit these activities.

Initially, 23andMe continued to sell the 'raw' genetic data (sequence of DNA letters without interpretation) and ancestry information. Outside the USA, the company continued to advertise and sell their services. In October 2015, the company announced that the FDA had given an authorization to market a more limited direct-to-consumer genetic test that included carrier status, wellness, trait and ancestry reports. Moreover, the product was launched in several countries outside the USA.

Also, not all the genetic factors have been mapped out. In order to do this, there is still a great need for scientific and epidemiologic research on large groups of patients and 'normal' individuals. Only then is it possible to develop algorithms that could, based on this huge amount of data, determine a certain risk. The risk of a genetic disorder is like the risk of ending up in a car crash: you do not know if or when it is going to happen you only know your chances are higher when you are a reckless driver.

More Pressure on Healthcare

Knowledge of your genetic profile may be presented as a basic right, but you still need experts to tell you what to do with the information. Even though most of these companies that offer genetic tests will tell you that their tests are not sufficient to make a diagnosis or any other medical decision, they do advise the consumer to take his or her test results to a doctor. This could possibly lead to higher consumption in healthcare services at a time when we need to save on our medical expenses. At this point the doctor also has to deal with a test that he cannot really estimate for value or significance. Although his medical knowledge is greater than his patients', the doctor is confronted with a test result that he did not ask for and has no idea how it was calculated. The chances are that this could lead to more tests and unnecessary interventions.

"Eating habits, weight, smoking behavior and so forth are better at predicting one's risk for developing heart and cardiovascular disorders.

And on the contrary, these tests can also lead to a false sense of assurance. Some test results will give you a low risk for a certain disorder. But genetic variations usually have a predictive value that is so small that people with a 'good' profile (for instance with heart and cardiovascular disorders) are not much better off in reality than people with a 'bad' profile. Eating habits, weight, smoke behavior and the likes are better at predicting one's risk for developing heart and cardiovascular disorders. It becomes really flagrant when someone is falsely given a sense of assurance with a test. Some test participants got the results that their risk of breast cancer was estimated as being very low. This conclusion was based on the absence of three variations in the BRCA1 and BRCA2 genes. They were three mutations that are found frequently in female breast cancer patients from the community of Ashkenazi Jews, but rarely in other women. Unfortunately there are known examples of hundreds of other mutations in the BRCA1 and BRCA2 genes that were not included in the test (one of the reasons is that the genes are patented, this will be explained in Chapter 22).

Are These Results Actually about Your Genetic Material?

A worst-case scenario would be that the laboratory accidently muddled your samples and that you have received someone else's test results. Things like that have happened. Fortunately those chances are small. Most of the laboratories meet quality standards and follow strict procedures to avoid errors like this as much as possible.

Examples exist of 'genome hackers' who have submitted biological samples without the knowledge and consent of someone. You could send in someone else's biological sample and then collect their test results. Anecdotal evidence also showed that some companies were willing to process a biological sample in order to offer genetic test results as a 'gift' to someone. This possible lack of genetic privacy forms a large ethical problem with these kinds of commercial genetic tests.

Moreover, research shows that most of the commercial companies that offer genetic tests accept samples from minors without a problem. This contrasts with the medical world where there is a generally accepted guideline to only test minors if it is for their benefit. This also considers the protection for the right of autonomy of the minor and his/her right to privacy as well as his/her right not to know. It becomes really worrisome when parents have their children tested for assumed hereditary risks to then change their lifestyle habits in an unreasonable fashion without the existence of a well-founded scientific and clinical basis.

Do Governments Allow These Commercial Tests?

A few years ago there was a boom in companies offering commercial genetic tests, mainly over the internet. Meanwhile it appears that the demand does not follow the supply: a small percentage of the population is aware that these tests even exist and only a fraction of those find the test interesting enough to order one. A number of companies stopped offering these tests. Other companies put the rifle on the other shoulder and ask that a doctor prescribe the test. This way they are looking to be recognized by – or work together with – the healthcare system.

In the meantime, more and more governments choose to interfere with online genetic tests for health purposes, or sometimes just completely forbid them. An example can be found in the case where the FDA, the American watchdog for food and health safety, prohibited the genetic test of 23andMe. Also countries like Germany, Portugal and France prohibit genetic tests without a doctor's prescription and guidance. At the European level negotiations are taking place to develop a new regulatory framework for genetic tests. At this time, France goes the furthest: it is not just commercial genetic tests that are forbidden; individual consumers that order such tests can be prosecuted and sentenced. However, enforcement might turn out to be very difficult for any limiting legislation, as the internet is a fluid medium that can be accessed from everywhere.

22 Genes and Patents

Genes occur naturally. The question remains if patenting them is a good idea. The current situation does cause some problems.

Genes can be patented. According to European and American patent laws everything that has been invented by human interference or can be isolated from nature can potentially be patented. You do have to be the first one in order to get your invention patented. As soon as you get the patent, you can determine who gets to use your invention and at what price. The system leads to progress – at least that is what Benjamin Franklin and Abraham Lincoln envisioned when they proposed the idea and turned it into law.

Patenting Genes, a Good Idea?

But in the case of a patent on a gene this principle is stuck: a gene exists only once, you cannot 'invent around it'. If you invent an electronic component for a phone or a fiber optic cable, your competitor can develop a different and better product. But a gene is a gene. And besides, it is a part of the human genome, which we all share in common. Yet, in recent decades, thousands of genes have been patented, with some roughly estimating about a quarter of our genes. In some cases the patent holders made lots of money; but in most cases the profit was not that great.

“A gene is a gene. And besides, it is a part of the human genome, which we all share in common.

In 2001 the European Patent Office granted the patent for the BRCA1 gene, which plays an important part in hereditary breast and ovarian cancer (see Chapter 5). European genetic centers and cancer research institutes protested against the decision. They defended the opinion that genes and also genetic defects that lead to hereditary disorders are elements from nature that cannot be patented. This is of the utmost importance, because otherwise performing genetic tests becomes impossible, even if their sole purpose is to trace genetic defects in patients and tie them to a risk of a hereditary disorder.

In total there were 3 patents on the BRCA1 gene and one on the BRCA2 gene. These patents, in a slightly modified way, survived the legal proceedings before the European Patent Office that lasted from 2001 until 2010. Strictly speaking the European genetic laboratories that conduct tests for BRCA1 and BRCA2 should get a license from the American company Myriad Genetics, which owns the patents. But to this day nothing has been paid. One reason is that the company is not interested in issuing licenses: it would rather test all the blood samples from Europe in its own laboratories and maintain the monopoly. Another reason is that you can only force the individual laboratories to cease their testing activities for BRCA1 and BRCA2 by means of many individual lawsuits in different countries, so the legal costs would probably exceed the profits. So that is probably why the company decided to open up a branch in Europe and currently offers the tests in a competitive manner. It looks like these tests are going to be available at below the market price so Myriad's competition, the public and academic laboratories, will be forced out of the picture, and then the company will raise its prices once it has an absolute monopoly. Genetic centers are not just going to accept that though. It is about the patient and the way public healthcare can keep offering the tests to those needing them, not who can afford them.

A Spanner in the Works of Patent Holders

The whole hypothesis that you can patent genes (according to the way they appear in nature on the condition that you can show that you have isolated them) was destroyed in 2013 by the U.S. Supreme Court. It determined that a piece of DNA from nature cannot be patented based on the fact that it was isolated and read. This was the verdict in the Myriad case about patents on breast cancer genes. In 2009, American associations of geneticists and pathologists together with some lobbyists also started lawsuits against similar patents in the United States. The case came before the Supreme Court in 2013 and was a serious blow for Myriad Genetics. The company and other pro-patent entities still claim that even though you cannot patent genes, you should still be able to patent genetic tests, despite the Court's decision. Luckily, a few subsequent US Court decisions have rejected patents on genetic testing.

> **"It is about the patient and the way public healthcare can keep offering the tests to those in need, not who can afford it.**

A court ruling in the United States changes nothing about European legislation, of course; but it does influence the way people think and act about patents on genes. Still there are pro-patent entities. In 2013 a Japanese company won a lawsuit in Germany against other companies that offered patented genetic tests without a license. The Japanese company owns patents on the FLT3 gene, which plays a role in certain forms of leukemia. This decision means that all the tests need to be conducted in the one German laboratory that is licensed to use the patent. The lab started acting aggressively and demanded that all blood samples from patients that need the test from different European genetic centers be sent to Germany. Something to which genetic centers answered in short and powerful manner: No way!

The Law: Clarity Wanted

Article 5 of 'Directive 98/44/EC of the European Parliament and of the Council of 6 July 1998 on the legal protection of biotechnological inventions' states the following:
"1. The human body, at the various stages of its formation and development, and the simple discovery of one of its elements, including the sequence or partial sequence of a gene, cannot constitute patentable inventions.
2. An element isolated from the human body or otherwise produced by means of a technical process, including the sequence or partial sequence of a gene, may constitute a patentable invention, even if the structure of that element is identical to that of a natural element.
3. The industrial application of a sequence or a partial sequence of a gene must be disclosed in the patent application."

You probably also had to read it twice and probably had to search for the difference between the first and second paragraphs. In the first paragraph it says you cannot patent genes as long are they are still within the human body. But as soon as they are isolated you can, even – and this is quoting the legislation – if the structure of that element is identical to that of a natural element.

This is some heavy writing, is it not? The Parliament needs to write these complicated sentences to say something without saying anything. The negotiations for this Directive took ten years. The Directive needs to be put in its context though: Europe stood idly by and watched biotech industry develop in the United States and wanted to use protective legislation to shift the tide. Strictly speaking this Directive was not necessary for patents on genes: through the eyes of the European patenting agencies, patenting genes was already possible based on the European patent laws dating from 1973.

Based on European law you can patent any invention as long as it is new, inventive, useful and described in detail. You cannot patent discoveries or ideas.

Patents and Complete Genome Analysis

So if around 5,000 of the 22,000 genes in our genome are patented, how can you still study a patient's genome in a reasonable fashion? You would basically need a license for each of the patented genes and to pay royalties to the patent holder. There has been a lot written on this topic, but the conclusion is simple: it is impossible, so the system of patenting genes and especially that of patenting genetic tests undermines itself. In any case this is the opinion of many geneticists that have been following this topic for all these years. Lawyers probably share a different opinion. Of course a good compromise is always welcome.

With the start of the Human Genome Project in the early 1990s, the discussions about genes and patents kicked off. The National Institute of Health (NIH) decided to patent all the DNA sequences that were collected as a result of the project. After a lot of debate and opposition they had to cancel their plans. The controversy was brought back to everyone's attention when the project was ending. Celera Genomics, Craig Venter's company (see Chapter 2), was trying to sequence the genome in one go and kept this information secret in order to try to patent as much of it as it could. In any case, this was Venter's plan. Not because of his personal conviction – at least that is what he writes in his book *A Life Decoded. My Genome: My Life* – but because of the financial backers of his company,

NEXT, a Visionary Book by Michael Crichton

In 2006 the successful American author Michael Crichton really cracked down on gene patenting in his book *NEXT*. 'Stop patenting genes,' he wrote as an addition to his novel. 'Genes are a fact of nature. Like gravity, sunlight and leaves on trees, genes exist in the natural world.' He compares the patenting of genes with the patenting of noses. Imagine that your nose was patented, then you could rub sunscreen all over your body but not your nose. You could no longer use makeup or sniff a delicious meal without paying the patent holder. 'Absurd,' as he calls it. And Myriad Genetics also gets a piece of his mind.

The book is really worth reading for those who love to mix fiction and science. Or the way Crichton puts it in his book: 'This novel is fiction, except for the parts that aren't.' Crichton also wrote for the television series *ER* and the books that formed the basis for the *Jurassic Park* franchise. He went to medical school, but is mainly active as a best-selling author, screenwriter and movie producer. He died unexpectedly in 2008. Fortunately he left a book that mentions DNA, genomes, syndromes, cloned apes, wacky researchers and even epigenetics.

namely a number of big pharmaceutical corporations. In the end, and as a result of the race between Venter's private genome initiative and the public international genome project, the company did not patent the genome as a whole. The sequence of the genome became public, but the possibility to patent individual genes still remained. One of the best-known adversaries of patenting genes is the Nobel Prize winner John Sulston. At the time of the Human Genome Project, he was the head of the English Sanger Institute and led the international branch of the genome project (i.e. the non-American contributions to the project). He did not receive his Nobel Prize for his contribution to the Human Genome Project, but for the research that preceded it. Its subject was a tiny nematode, *Caenorhabditis elegans*. This organism, only a millimeter long, is a model organism for the study of many cell biological processes.

The Case of Myriad Genetics

The best-known discussion about genes and patents that was widely covered by the media was the one about the BRCA1 and BRCA2 genes. During the search for breast cancer genes, many organizations formed an international consortium called the Breast Cancer Linkage Consortium (BCLC) to exchange material and results. In the last strait before the finish, one of the groups started an early sprint. Mark Skolnick, a researcher from the University of Utah, decided to go his own way. With money from the pharmaceutical and biotech industry, he founded Myriad Genetics, a company that would employ over 60 researchers to map out the part of chromosome 17 that should hold the BRCA1 gene. In 1994 they discovered a gene in which mutations were found in patients from breast cancer families. This had to be the BRCA1 gene. The company immediately applied for a patent with the American and European patent offices.

This was and still is no joke to Mary-Claire King, as well as to many researchers of the consortium, who were also on the gene's trail and watched the main prize get stolen right in front of them. Gilbert Lenoir, a French cancer researcher who was also in the race, described the researchers' disappointment like this: the whole group of researchers started to climb to the top, and in the last mile, one team took the lead. They were good climbers and made it to the top. But they only succeeded because the others had already brought them

Patents and Non-Invasive Prenatal Tests

There are a lot of patent protections when it comes to techniques that are used for the non-invasive screening of Down syndrome (see Chapter 8). Simply put, there are two kinds of patents. There is the original patent on the discovery of cell-free fetal DNA in the mother's blood. The fact that this is considered a discovery indicates that this patent has led to great discussion. Discoveries cannot be patented. There are a number of technical patents on the methods used to trace this cell-free fetal DNA. The later patents are not heavily criticized because it is always possible to invent newer techniques and thus create progress. But if cell-free fetal DNA remains fundamentally patented as a principle, there is a new problem that arises, similar to the problem with patenting genes: you cannot go around it, because they are natural phenomena and principles.

close to it. When they were descending after the climb, they made clear to the others that they were taking the prize money and were going to monopolize the summit. That really hurt the others.

But legally there was nothing that could be done: there were no agreements within the consortium about the intellectual property of the results and there were no signed or even drafted contracts. There is no one who knows what others from within the consortium would have done if they had been the first to discover the gene. That remains an open question...

A year later, something similar happened with the BRCA2 gene. Researchers at the Institute of Cancer Research in Cambridge (UK) discovered a part of the gene and applied for a patent – they say it was to protect themselves against Myriad Genetics, which is probably true. But because Myriad was the first to isolate the entire gene, it defeated the grant of a patent on the whole DNA sequence of BRCA2.

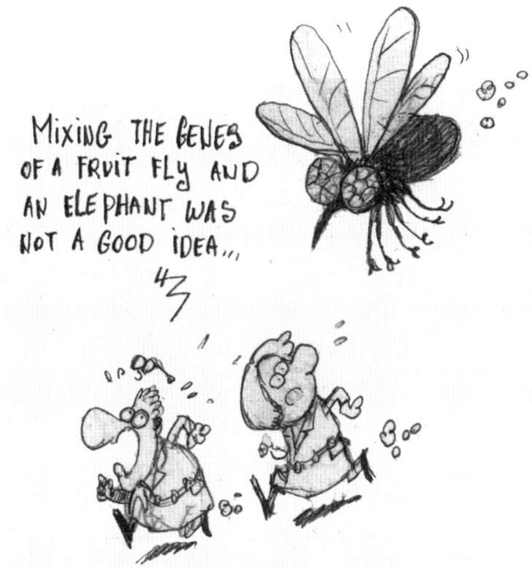

Misappropriation of Patents

A patent mainly offers the ability to prevent someone from using and commercializing an invention. Myriad Genetics – and with it the University of Utah and a research group of the National Institutes of Health (NIH) from North Carolina as the patent's co-holders – decided not to offer licenses for its patent. Myriad acquired the monopoly for a genetic test. This test determines the patient's DNA sequence of the BRCA1 and BRCA2 genes, and compares it against the reference sequence. Mistakes in the gene, which are predictive of whether or not the function of the respective protein will be disturbed, are considered the cause of the higher cancer risk, and thus the cause of the disorder in the patient and the family.

> **“A patent mainly offers the ability to prevent someone from using and commercializing an invention.**

As a result of the patent and the legal threats it brings, all genetic laboratories and research institutes in the US stopped their research into BRCA1 and BRCA2 mutations in patients as of 1996 – the year the patent was approved by the American patent office. Myriad Genetics acquired a complete monopoly. Even Angelina Jolie had to rely on Myriad's laboratory for her test. The price of the test is somewhere between 2,000 and 3,000 USD. In Europe, genetic centers and cancer institutes kept offering the test until it became known in 2001 that the patent had been granted in Europe as well. Instead of complying with the monopoly, European laboratories started taking legal action against the patent. They never caved in to the threat and continued offering the BRCA1 and BRCA2 tests, even though conducting the test without a license is illegal, strictly speaking.

> **“Money and healthcare, they coexist in a weird way.**

Myriad Genetics has since put its rifle on its other shoulder. Up to now the company has not started any lawsuits against laboratories in Europe. It could have done this because patent rights partially stayed intact. But

23andMe's Controversial Patent

In the fall of 2013 the personal genomics company 23andMe registered a patent for a method of analyzing genomic data and making predictions about how the offspring of a random couple could look. In the patent, 23andMe describes the possibility to compare genetic material from different donors and pick the donor that offers the biggest chance for a desired trait in your offspring. In short: if a woman wants to become pregnant and she has to rely on a sperm donor, this technology should enable her to pick between different donors and choose the one that offers the greatest chance for a child to have a desired trait or not to have a certain trait or condition. In the patent 23andMe also describes different traits that can be used as tools of selection, including eye color, muscle development, personality traits and even different multifactorial disorders. 23andMe is also quick to say that this is just a statistical calculation and that the company cannot come to a conclusion about how the offspring will actually look. This patent comes very close to the idea of 'designer babies'. The use of genetic diagnostics in pre-implantation to select embryos that do not show certain hereditary disorders is already a topic of many different ethical discussions (see Chapter 9). It is clear that a computer program made to select the best donor based on a desired trait only adds fuel to the ethical discussions surrounding designer babies. Yet the patent was described in this manner. For the moment the company is offering this technology for 'recreational purposes'. It gives individuals the ability to compare their genomic material with that of other individuals and makes predictions about the traits their virtual children could have.

the company clearly prefers a commercial battle to a legal one: Myriad started a service company in Germany and now offers tests on a commercial basis. Since the patents expired in 2015 this was probably the smart thing to do.

Besides, these centers made the transition to the new generation of DNA sequencers before Myriad did, which allows them to conduct the tests more cheaply and faster than before. The centers are not giving up the BRCA tests. It would be disastrous for patients or healthcare systems if they had to pay more. But Myriad has deep pockets, partly thanks to the huge profits it has made with these tests in the U.S. In 2009, Myriad, according to its own tax declarations (Myriad Genetics is a public company), generated a 329 million USD revenue from tests that cost 43 million USD, a great profit margin that is largely due to the monopoly. There have also been rumors that representatives from the company told a number of French cancer institutes that they are ready to drop their prices in Europe until all the public laboratories have to give up the tests. After that they will have the market to themselves and they will increase prices. Money and healthcare: they coexist in a weird way.

23
X and Y

It is not entirely true that all people have 23 pairs of chromosomes. Only women have two X chromosomes, men have an X and a Y chromosome.

We each have 23 pairs of chromosomes. Each pair has one chromosome from the father and one from the mother. A man does not have two equal sex chromosomes: he has an X and a Y chromosome, whereas a woman has two X chromosomes. In reproduction, a woman will always pass an X chromosome, whereas a man will pass either an X or a Y chromosome. The Y chromosome determines the sex of the child. In other words: it is the father who determines the child's sex. Always? Always!

From a Genetic Man to a Real Man

The Y chromosome must hold something special to turn a genetic man into a real one. And yet, in Chapter 1 we mentioned that the Y chromosome barely contains any genes, around fifty, especially compared to the other chromosomes and especially with the X chromosome, which contains around 1,000. Researchers showed that it does not take a lot to turn a woman into a man: 1 gene, but it is an extraordinary one!

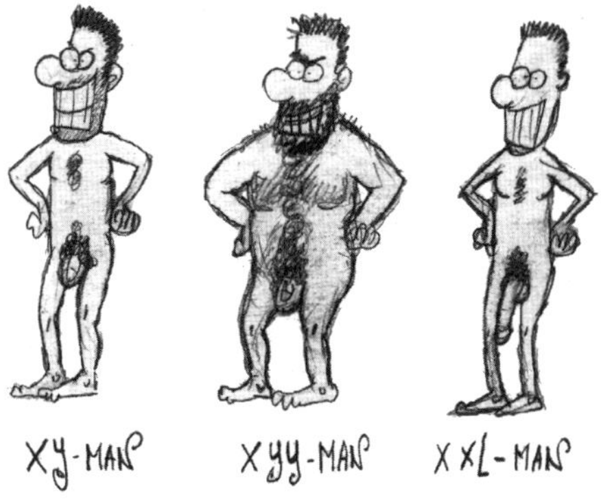

The SRY gene is a relatively small gene that is present only on the Y chromosome and does not have a counterpart on the X chromosome. This gene codes for a protein that can bind DNA and can thus give the cell instructions for turning other genes on or off. There are many different proteins that have a similar function in the cells, the so-called 'transcription factors': they determine whether or not a specific gene in a specific cell will be activated in a specific circumstance. In other words, they determine if at that specific time, in those circumstances you need that recipe.

> **“Sometimes an individual's appearance will not match the appearance predicted by the chromosomes.**

The SRY protein does this at a crucial time during the embryo's development. The protein gives an important command to the cells in the embryo that are planning on constructing a female reproductive organ. Something along the lines of, 'Hey, let's take a different approach. Let's create some testicles.' To say it in a more scientific manner: each embryo builds the foundations for the genitalia at an early stage, but only during the seventh week of pregnancy are there hormones emitted that will direct the growth in either the male or female direction.

If the SRY protein is not present, the embryo will become female *by default*. If the SRY protein is present, it will spark the creation of testicles that will produce the hormones needed for the further construction of male reproductive organs. Some conclude from this that every man is in essence a woman who is reprogrammed and rewired during a specific time in the embryonic development, thanks to the SRY gene.

That the SRY is important and unique was proven in 1991 by implanting the sry gene (note that the human genes are capitalized but the corresponding animal ones are not; this is an international agreement) from a male mouse into a female mouse. The mouse started showing masculine traits. Did the mouse immediately start drinking beer, the average man asks himself. No, it is not about that. It is about the construction of the primary sex characteristics.

Like we said earlier, there are not many genes on the Y chromosome, not even all of the genes a man needs. Take for example the genes that play a part in the production of testosterone. These are scattered over the genome. Men and women both have these. Everything depends on how their use is regulated.

There are some genes on the Y chromosome that are important for spermatogenesis. We know this because men who are missing a piece of their Y chromosome (a piece that does not contain the SRY gene otherwise they would not be men and would not have primary reproductive organs) have fertility problems. They suffer from azoospermia: they cannot ejaculate sperm cells because those do not 'mature' in the testicles. They do have the sperm cells' progenitors in their testicles. With an 'intracellular sperm cell injection' (ICSI) it is possible for these men to become fathers. ICSI is a sophisticated form of in vitro fertilization (IVF) in which the nucleus (and thus the DNA) of the (premature) sperm cell is inserted into an egg cell through a micro-injection, after which the fertilized egg cell is implanted into the mother. The (premature) sperm cell is harvested through testicular biopsy. It is important to note that boys born after this procedure will have the same problem as their father because they inherit the same Y chromosome lacking the piece of DNA.

Sex Reversal

Sometimes an individual's appearance will not match the appearance predicted by the chromosomes. There are people who do not have a visible Y chromosome, yet they do have male reproductive traits. They are called 46,XX males (46,XX is the simple way of writing the human chromosomal pallet: 46 chromosomes, of which two are X chromosomes). 46,XX has nothing to do with hormones, but everything to do with the SRY gene. Because of an unhappy exchange of genes between the X and the Y chromosomes in the father, the SRY gene sometimes lands on the X chromosome. This jump does not deactivate the gene, allowing it to do its normal task: telling the embryo that it should start developing male reproductive parts. These people are genetically female but have the primary reproductive traits of a man because they carry the SRY gene on one of their two X chromosomes. Unfortunately the change does not go smoothly since the rest of the Y chromosome is lacking. Around 80% of the time after they have hit puberty, they will have normal pubic hair and a normal penis and will feel like a man. But they are sterile because of a lack of normal sperm cells. They also develop breasts and are therefore treated with testosterone. The other 20% of

All Men in the Family

Every man's male descendants, who carry his family name, (usually) carry his Y chromosome. So the son of the son of the son, etc., has the same Y chromosome as his great great-, great-, etc., grandfather.

In 2013 there was a book published about the search by two genealogists for all the descendants of a certain Joos and Jacques Cassiman who lived in the Flemish village of Lebbeke around 1600. They mapped out more than 7,000 descendants spread all over Belgium. Jean-Jacques Cassiman, a geneticist at the University of Leuven is one of them. He conducted genetic research into the Y chromosome of a number of Cassi-'men'. They all seemed to have the same Y chromosome, except for a couple of typos that managed to find a way in over the generations. This confirms that they all descend from the same forefather.

the time 46,XX men are born with ambiguous genitalia and will have more obvious problems.

> **"An extra X or Y chromosome has no or mild clinical symptoms.**

Something else happens with 46,XY women: they lack the SRY gene, even though their Y chromosome looks intact under a microscope. In most cases the gene is absent or a mutation has caused it to turn off. The genetically male individuals will not develop the primary male genitalia. Despite the presence of a Y chromosome, they will have the physical appearance of a woman. In 2009 the press suggested that this was the case with the South African Caster Semenya, the female world champion in the 800m. Some papers even used the word hermaphrodite, which is completely out of place: this term is not used when talking about humans. Fortunately the result of the genetic test was never made public; it is only the athlete's business. Because of this occurrence the IAAF decided that testosterone values are important for the admission procedure of the Olympics, not the outcome of a genetic test.

Adultery...

Following a Y chromosome throughout the centuries also reveals a lot of information about 'who the father is' or 'was'. In 2013 researchers from Leuven studied the inheritance of genetic markers on the Y chromosome based on the family name for a large number of families. What was the study's frame? Checking how much evidence there is for *extra-pair paternity*. The sociological data about this topic is limited and that is why the historical study is useful. By checking the Y chromosome of different people with the same name throughout large family trees, researchers found that the percentage of extra-pair paternity throughout the centuries was around 1 to 2%. That is a lot less than the 8 to 30% suggested by some behavior studies. It is close to what other studies in Western Europe tell us. This also means that most family trees are correct and that genealogists do not have much to worry about when they map out their families.

It debunks a myth that often appears in newspapers – and is kept newsworthy by (internet) companies that offer paternity tests – namely that almost 10% of all children are not conceived by their mother's marital partner. In (historical) Flanders (Belgium), at least, this seems not to be the case.

In both cases the genetic disorders are rare. There are other causes of *sex reversal*, making the clinical diagnosis more complex than is described in this book.

Too Much X or Y, a Problem?

In Chapter 4 we mentioned that numerical chromosomal abnormalities, generally speaking, have severe consequences. Most are not viable or cause noticeable disorders like Trisomy 21, which leads to Down syndrome. The sex chromosomes form the exception to this rule. An extra X or Y chromosome has no or mild clinical symptoms. The following chromosomal anomalies are found with a frequency of approximately 1 every 1,000.

XYY

A man with two Y chromosomes. Half a century ago researchers published a study that noticed that in prison relatively more 47,XYY males could be found than in the average population. The results of this study were never confirmed but its message was a hard one to get rid of. The alternative explanation that was put forward is that XYY is associated with some degree of cognitive impairment, suggesting that prisons have more XYYs because XYY criminals are more likely to be caught than XY criminals, but not more likely to be criminals in the first place. There is not a single scientific or biologic piece of evidence that an extra Y chromosome would lead to aggression or super-hero-like traits. On the contrary: just YY, so without the X chromosome would not be viable. Reason: there are too many essential genes on the X chromosome.

XXX

These are triple X women. People with a 47,XXX chromosome pallet are still female and are hardly noticeable, not even medically. They are generally a bit taller or more fat but this is not exceptional. They are fertile because they can produce normal egg cells.

XXY

This genetic defect has a name: Klinefelter's syndrome. These are people with a manlike appearance, but they have a number of disorders that usually escape even diagnostics. They are often taller than average, with remarkably long arms and legs. They often, but not always, have small testicles. They are infertile and oftentimes the diagnosis is made because of the patient wanting a child and having fertility problems. XXY-men usually have normal intelligence with an IQ that is slightly lower than their XY sibs, and they can sometimes suffer from behavioral and learning problems. There are Klinefelter variations that come with even more extra X chromosomes: 48,XXXY or 49,XXXXY, but these are extremely rare. The symptoms are similar but they also come with a lower IQ and more severe behavior issues as the number of extra X chromosomes increases.

X-Linked Disorders and X-Linked Inheritance

The X chromosome contains approximately 1,000 genes, and in theory there should be as many hereditary disorders related to the X chromosome. The X-linked disorders follow a number of inheritance laws that are pretty special. To remember them, you might want to

Color Blindness

Color blindness or color vision deficiency is actually the unifying terminology for different deficiencies. A specific form of color blindness, red-green deficiency (problems with distinguishing between red and green) is clearly determined genetically: 1 in 12 (8%) men has it, but only 1 in 200 (0.5%) women. The OPN1MW gene that codes for the green pigment in the cones in the eyes lies on the X chromosome and any defects are recessive. Women have two X chromosomes, so the odds of the recessive gene being present on both chromosomes at the same time are small. Men have only one X chromosome; as soon as they inherit the altered gene they are toast.

Yellow-blue deficiency is also hereditary but the gene that causes it is based on a different chromosome that is present twice in men and women. This form of color blindness is present equally in males and females – or, better put: in 0,0001% of the earth's population.

take a closer look at the next two remarkable phenomena associated with this sex chromosome. First of all - and this is the most important part - X-linked disorders affect boys more often than girls because boys have only one X chromosome. Boys that have a genetic defect on their X chromosome do not have a *back-up* in the form of a second X chromosome that could have the healthy gene. For girls it is the opposite: they have the *back-up* on the other chromosome. Girls will rarely show symptoms of X-linked (recessive) disorders. This is why it is important to prevent these disorders in families where they occur, by figuring out which women are carriers. They have a 50% chance of having a boy with the disorder.

Second of all: X-linked families, as they are called, can be recognized through their family tree. The boys have the disorder, girls can be carriers, but you will never see a father and a son with the same clinical problem since fathers never pass their X chromosomes onto their sons. The result is that for many disorders, like hemophilia, it has been known that they are X-linked for a long time. Some say the ancient Greeks already figured this out. This led to genetic research and the search for genes on the X chromosome starting earlier and with a higher success rate.

It allowed researchers to limit them to a gene hunt on the X chromosome without having to worry about the rest of the genome. The list of X-linked disorders is very long. Fabry's disease and Adrenoleukodystrophy, two neurodegenerative disorders; Alport's syndrome, a kidney disease; Coffin-Lowry's syndrome; Barth's syndrome; Hunter, Lowe, Lesch-Nyhan, Menkes disease, Wiskott-Aldrich, are only a few of the X-linked disorders that have been known for a long time. Many of the X-linked disorders result in intellectual disability. The best example is the fragile X syndrome, which can account for approximately 1% of all intellectual disabilities in boys, and which is linked to the FMR1 gene. Moreover, there are hundreds of other genetic defects on the X chromosome that typically cause intellectual disability in males. It is important to determine the precise cause of an X-linked condition leading to intellectual disability in order to be able to trace potential carriers of the disorder.

For completeness, it should be mentioned that there are also dominant X-linked disorders. In that case the girls will show the symptoms and, sometimes, the boys that would be affected are not even born because of the severity of the defect. The Rett syndrome qualifies for this category. It is a disorder that affects young girls. After what appears to be a normal early development for approximately a year, girls with Rett syndrome will start showing a standstill or deterioration in their development. Things they could do before, like talking and walking, they can no longer do, and they start living in their own world. The stereotypical hand gestures and a very strong eye contact are typical. Some girls also develop epilepsy and orthopedic deformities. Rett syndrome is principally a result of a 'de novo' mutation (see Chapter 4).

A man never passes his X chromosome to his son. This is why an X-linked hereditary disorder cannot be passed from father to son. This is a useful observation to exclude an X-linked disorder in a family. However, watch out, a father gives his X chromosome to each of his daughters. All daughters of a man with an X-linked disorder become carriers and their sons have a 1/2 chance of inheriting the disorder from the grandfather. X-linked inheritance can be seen in the family tree of the British Queen Victoria: her youngest son Leopold is the only one that had hemophilia; her daughter Alice was an obligate carrier, and that became clear at the time her son Rupert was born with hemophilia.

Women Biologically Stronger? X Inactivation

Women have two X chromosomes; men have only one. How is it possible that women are not bothered by the 'overdose' of genes that is present on their chromosomes? Or the other way around: how is it possible that men are (relatively) normal with only one X chromosome? The answer is that one or both X chromosomes in every cell of a woman is inactivated, the so-called 'X inactivation'.

This is an interesting phenomenon biologically speaking because it balances the equilibrium. The X inactivation happens through the chemical modification of the DNA comparable to the epigenetic phenomena mentioned in Chapter 4. In extraordinary cases, an altered X inactivation also has clinical consequences. Those can lead to carriers of X-linked disorders showing symptoms of the disorder that are expected to be seen only with boys, like hemophilia A or B, or Duchenne disease.

Too Crazy

Collaborators at the genetic center in Leuven (Belgium) received two requests to research the 'royal' hemophilia B mutation. The specific mutation that was present in the offspring of Queen Victoria (who ruled over the United Kingdom from 1837 to 1901), is the C.278-3A>G mutation in the factor 9 gene (F9). There are no known living descendants of the Queen who carry the mutation. The mutation was found after genetic research on the remains of Queen Victoria's great grandson, the Russian Tsarevich Alexei (see Chapter 16). In less than a year, two women consulted with the Leuven-based geneticists with a request to perform very specific genetic research – which is weird in a case like that. Hemophilia B can be caused by any possible mutation in the factor 9 gene. In both cases the women claimed that they had sons with hemophilia, but that their sons did not want to come with them to the consultation. Something they made up? Probably, but if the plan works you have proven that you have got English Royal blood flowing through your veins. Maybe this suggestion circulates somewhere on the internet, or in high society circles or ladies' clubs? We do not know, but we did not let them outflank us, so new candidates need not apply.

Recommended Literature

Books

Collins, F., *The language of life: DNA and the revolution in Personalized Medicine*, New York, Harper Collins, 2010.

Davies, Kevin, Cracking the genome: Inside the race to unlock human DNA, New York, Free Press, 2001.

Henderson, Mark, 50 Genetics Ideas you really need to know, Quercus Editions, London, 2008.

Jones, Steve, The language of genes. Biology, history and the evolutionary future, Flamingo, London, 1993.

Lone, Frank, *My beautiful genome: exposing our genetic future, one quirk at a time*, Oneworld Publications, Oxford, 2011.

Sandel M., *The case against perfection: Ethics in the age of genetic engineering*, Cambridge, Mass., Belknap Press of Harvard University Press, 2007.

Stone, Linda & Lurquin, Paul F., *A genetic and cultural odyssey. The life and work of L. Luca Cavalli-Sforza*, New York, Columbia University Press, 2005.

Venter, J. Craig, *A Life Decoded. My Genome: My Life*, Penguin Books Ltd, 2007

Articles and Reports

Cornel MC, Rigter T, Weinreich SS, Burgard P, Hoffmann GF, Lindner M et al., *A framework to start the debate on neonatal screening policies in the EU: an Expert Opinion Document*, Eur J Hum Genet 2014, 22(1):12-17.

European Society of Human Genetics, *Statement of the ESHG on direct-to-consumer genetic testing for health-related purposes*, Eur J Hum Genet 2010, 18(12):1271-1273.

Expert Group on Dealing with Ethical and Regulatory Challenges of International Biobank Research, *Biobanks for Europe. A challenge for governance*, Brussels, European Commission, Directorate-General for Research and Innovation, 2012.

Health Council of the Netherlands. Screening: between hope and hype. The Hague: Health Council of the Netherlands, 2008; publication no. 2008/05E. ISBN 978-90-5549-722-5

Health Council of the Netherlands. The 'thousand-dollar genome': an ethical exploration. Monitoring Report Ethics and Health, 2010/2. The Hague: Centre for Ethics and Health, 2010. Publicatienumber Health Council of the Netherlands: 2010/15E. ISBN 978-90-78823-16-2

Health Council of the Netherlands. NIPT: dynamics and ethics of prenatal screening. The Hague: Health Council of the Netherlands, 2013; publication no. 2013/34. ISBN 978-90-5549-983-0

Green RC, Berg JS, Grody WW, Kalia SS, Korf BR, Martin CL et al., *ACMG recommendations for reporting of incidental findings in clinical exome and genome sequencing*, Genet Med 2013, 15(7):565-574.

Human Genetics Commission, *Inside information. Balancing interests in the use of personal genetic data*, London, Human Genetics Commission, 2002.

Human Genetics Commission, *Profiling the newborn: a prospective gene technology*, London, Human Genetics Commission, 2005.

Human Genetics Commission, *Making babies: reproductive decisions and genetic technologies*, London, Human Genetics Commission, 2006.

McNally E, Cambon-Thomsen A., *Ethical, legal and social aspects of genetic testing: research, development and clinical applications*, Brussels, European Commission, 2004.

Nuffield Council on Bioethics, *Genetics and the human behaviour: the ethical context*, 2002.

Nuffield Council on Bioethics, *Pharmacogenetics: ethical issues*, 2003.

Nuffield Council on Bioethics, *The forensic use of bioinformation: ethical issues*, 2007

Nuffield Council on Bioethics, *Medical profiling and online medicine: the ethics of 'personalised healthcare' in a consumer age*, 2010.

Nuffield Council on Bioethics, *Donor conception: ethical aspects of information sharing*, 2013.

President's Council on Bioethics, *The changing moral focus of newborn screening. An ethical analysis by the President's Council on Bioethics*, Washington DC, 2008.

Presidential Commission for the Study of Bioethical Issues, *Privacy and Progress in Whole Genome Sequencing*, Washington DC, 2013.

Presidential Commission for the Study of Bioethical Issues, *Anticipate and Communicate: Ethical Management of Incidental and Secondary Findings in the Clinical, Research, and Direct-to-Consumer Contexts*, Washington DC, 2013.

Public Health Genomics Foundation, *Next steps in the sequence. The implications of whole genome sequencing for health in the U.K*, 2011.

Public Health Genomics Foundation, *Public health in an era of genome-based and personalized medicine*, 2010.

Royal College of Physicians, Royal College of Pathologists & British Society for Human Genetics, *Consent and confidentiality in clinical genetic practice: guidance on genetic testing and sharing genetic information, 2nd edition*, Report of the Joint Committee on Medical Genetics, London, RCP, RCPath, 2011.

Secretary's Advisory Committee on Genetics, Health, and Society, *Realizing the Potential of Pharmacogenomics: Opportunities and Challenges*, 2008.

Secretary's Advisory Committee on Genetics, Health, and Society, *Gene Patents and Licensing Practices and Their Impact on Patient Access to Genetic Tests*, 2010.

van El CG, Cornel MC, Borry P, Hastings RJ, Fellmann F, Hodgson SV et al., *Whole-genome sequencing in health care. Recommendations of the European Society of Human Genetics*, Eur J Hum Genet 2013, 21 Suppl 1:S1-S5.

About the Authors

Pascal Borry is associate professor of bioethics at the Centre for Biomedical Ethics and Law (University of Leuven, Belgium). His research is focused on fundamental and applied research in bioethics. His main research activities are concentrated on the ethical, legal and social implications of genetics and genomics. He has published amongst other issues on topics such as direct to consumer genetic testing, public health genomics, biobanking, research on human tissue, genetic testing, preconception screening and neonatal screening. He has published as well on the relationship between empirical and normative approaches in bioethics. He has published more than 100 publications in international peer reviewed journals.

He is programme director of the Master of Bioethics at the University of Leuven and teaches medical ethics to medical students. He is a member of the Flemish Commission on neonatal screening (2012-2017), a member of the Belgian Consultative Committee on Bioethics (2014-2018) and expert of the Superior Health Council (2014-2020). Within the European Society of Human Genetics he was a member of the Professional and Public Policy Committee (2009-2016) and is elected member of the board (2012-2017). Within the university he was a member of the steering group YouReCa (Young Researchers Careers). In 2016 he became a member of the Ethics Panel of the World Anti-Doping Agency.

He has received various prizes. In 2006 he received the triennal prize for biomedical ethics 'Professor Roger Borghgraef'. In 2014 he also received the Innovation Prize of the Dutch Association for Community Genetics and Public Health Genomics, and in 2015 the prize of the Dutch Society for Bioethics. In 2015 he also received the Science Communication Award of the Royal Flemish Academy of Belgium for Science and the Arts. He was a visiting scholar at the Case Western Reserve University, the Université de Montréal, and the VU Medical Center Amsterdam. He is still Invited scholar at the Center for Genomics and Policy of Mc Gill University.

Gert Matthijs is head of the DNA laboratory at the Center for Human Genetics at the University Hospital in Leuven, and professor at the University of Leuven (Belgium). As a molecular geneticist, he has been involved in the diagnostics of inherited diseases since 1994. His research is on a group of rare, inborn diseases called Congenital Disorders of Glycosylation (CDG). He cloned the gene for the first and most frequent type of CDG, back in 1996. Since then, his team has contributed to the identification and elucidation of a large number of defects. He and his colleagues from the EUROGLYCANET network received the "Körber European Science Award" in 2004, for their contribution to the clinical and basic research on CDG.

His other research activities deal with the development and validation of novel technologies for genetic testing. His lab is using massively parallel sequencing for research and diagnostics. He has been the co-ordinator of EuroGentest, a network for quality and harmonization of genetic services in Europe. The network was originally funded by the European Commission as a project, and is now part of the European Society of Human Genetics (ESHG).

Together with his colleagues in genetics, he has coordinated a workgroup on legal, ethical and societal aspects of "total genome analysis" (in the context of the Metaforum initiative of the university). This lead to the book 'Wat met genetica?' that was published in 2013, and edited in French as 'Mes gènes, mon identité' in 2014.

From 2001 to 2010, he has spent a lot of time on leading the European opposition against the granting of patents on BRCA1 and BRCA2. This has been a wonderful experience, and a personal challenge, whereby he has tried to go beyond the Myriad case. As a result, he has published a lot about genes and patenting.

At the international policy level, he is a member of the Scientific Diagnostic Committee of IRDiRC (International Rare Disease Research Consortium) and an alternate member of the European Commission's Expert Group on Rare Diseases (formally EUCERD).

About the Cartoonist

Joris Snaet holds a PhD in History of Art and is an architectural historian of the KU Leuven Technical Services Department. He draws cartoons for the Belgian newspaper *De Standaard* and the university magazine *Campuskrant*. He has also drawn cartoons for *Science*.